DAMMING THE DELAWARE

DAMMING THE DELAWARE
The Rise and Fall of Tocks Island Dam

Richard C. Albert
Foreword by Stewart L. Udall

THE PENNSYLVANIA STATE UNIVERSITY PRESS
University Park and London

Library of Congress Cataloging-in-Publication Data

Albert, Richard C.
Damming the Delaware.

Bibliography: p.
Includes index.
1. Tocks Island Reservoir Project. I. Title.
TC425.D3A65   1987   333.91′62′09749   86-43197
ISBN 0-271-00481-9

To my parents
with gratitude and love

Contents

CONTENTS

Foreword

The stream known as the Delaware was the subject of a historic 1930s lawsuit that inspired Justice Oliver Wendell Holmes to write the oft-quoted words, "A river is more than an amenity, it is a treasure."

Damming the Delaware is a lively account of a graphic episode in our conservation history. It is the story of a decade-long fight over the fate of a river that took on national significance by becoming a struggle that dramatized the evolution of the environmental movement in this country.

My involvement in this controversy began in 1961 when I helped the governors of New York, New Jersey, Pennsylvania, and Delaware establish a singular federal-state institution called the Delaware River Basin Commission. I then served for seven years as the federal member of the DRBC, and played a lead role in setting the stage for the Tocks Island struggle when I urged Congress to authorize the U.S. Army's Corps of Engineers to construct a huge dam on a scenic stretch of the beautiful river.

What made the Tocks Island fight a classic conservation dispute—and what makes the details of this book so engrossing—is that it coincided with a turning point in our nation's outlook toward its resources. As the 1960s began, there was little dispute over this dam on the Delaware, but before that decade ran its course, the environmental movement had formed its phalanxes and was generating new attitudes and values that reversed the pro-dam policies nurtured for a half century by the leaders of both political parties.

This book should have a broad appeal because it is a case study that reveals,

with remarkable clarity, the dynamics of environmental reform in this country. The battle over Tocks Island Dam instructs us because it exemplifies how
social, ecological, economic, and legal issues became intertwined with conservation politics during the formative years of the environmental movement.

This environmental struggle is fascinating because it was a pitched battle
that encompassed a high dam that was stopped a few days before a scheduled
ground-breaking ceremony, a "war zone" created by the land-acquisition
fiascos of the Corps of Engineers, a "colony" of squatters led by a Squatters
Parents Committee, a lone Lenni Lenape Indian who appeared at hearings to
urge that the feds return a part of their newly acquired land to the natives, a
memorable hike led by Justice William O. Douglas, an environmental lawsuit
with an unusually large number of plaintiffs, one of the trail-blazing environmental impact statements prepared after Congress enacted the National Environmental Policy Act in 1969—and unceasing maneuvers by coalitions of
supporters and opponents of Tocks Island Dam.

In short, this is a story of a historic turnaround for many policymakers. I
offer myself as a case in point. The same year (1962) that I urged Congress to
authorize Tocks Island Dam, I also initiated the planning which later resulted
in the federal legislation establishing a National System of Wild and Scenic
Rivers. And today the main stem of the Delaware remains undammed. The
river that Tocks Island Dam would have inundated is now, by Act of Congress, a unit of our National System of Wild and Scenic Rivers.

Stewart L. Udall
February 1987

Preface and Acknowledgments

The Delaware River is one of the last undammed major rivers in the United States. The lack of a dam on the Delaware River is a marvel since the history of the Delaware centers on man's desire to dam it, culminating in the authorization of Tocks Island Dam in 1962. The dam got within weeks of ground breaking, but it has not been built. It remains the most controversial water project in the Delaware Valley and, perhaps, the entire East.

It has been my objective to write a comprehensive, historical account of the Delaware and its primary dam project, Tocks Island Dam. Although millions of words have been written on these subjects, the evolutionary process has never been described in any detail. It is this evolutionary process that largely governs water management in the Delaware River Basin today.

The Delaware River/Tocks Island story was derived from extensive research of the public record and through interviews. The sheer volume of this material is unbelievably large and complex—and usually biased. It has been my goal to cull this material and to present it in a rational, neutral fashion. As the result of my selection process, I am sure some readers may object to what was emphasized or, conversely, downplayed or ignored. I offer no apology except to recognize that many subjects touched on in this book could have been explored in greater detail. A subject for which I would like to see additional professional study is the striking parallels between Tocks Island Dam and the Vietnam War.

During the course of researching this book, I visited many libraries. The

Delaware River Basin Commission has the single best collection of Delaware River material that I found. The DRBC library contains materials written since its founding in 1962, plus material from the Interstate Commission on the Delaware River Basin and the Delaware River Basin Advisory Committee. The commission's news-clipping files, maintained by the Public Information Office, were also particularly helpful. The New Jersey State Library has a collection of material that complements the DRBC collection, plus an especially helpful and competent library staff.

I would like to thank the many persons who helped me in one way or another during the preparation of this book. Special thanks go to Dorothy Belmont, Bev Conover, Tom Detweiler, Bob Horton, Tom Iezzi, Gretchen Leahy, Joan Matheson, Bill Read, Nancy Shukaitis, Paul Webber, Sue Weisman, and Buzz Whitall. The contents of this book are, however, my own responsibility.

Finally, I would like to express appreciation to my wife, Mary, and daughters, Wendy and Carrie, for their extreme patience during the long course of this project.

Richard C. Albert

Delaware River Basin Facts

Drainage Area: 12,765 square miles (0.4 percent of U.S. land area), with 6,780 sq. mi. at Trenton.

Length: 331 miles, including Delaware Bay with 200 miles of nontidal river (Hancock, NY, to Trenton, NJ).

Average Discharge of the Delaware River at Trenton: 11,750 cubic feet per second (33rd largest U.S. river in terms of flow). Maximum flow was 329,000 cfs on August 20, 1955. Minimum flow was 1,180 cfs on October 31, 1961. Typical summer low-flow is controlled at 3,000 or higher.

Population: 7 million (a population greater than most states), with a service area of 20 million (almost 10 percent of U.S. population).

Average Rainfall: 44 inches (about 10 trillion gallons).

Miles in National Wild and Scenic Rivers System: 120 (one of the last major U.S. rivers without a dam on its main stem).

Largest Tributaries: Schuylkill River (first) and Lehigh River (second).

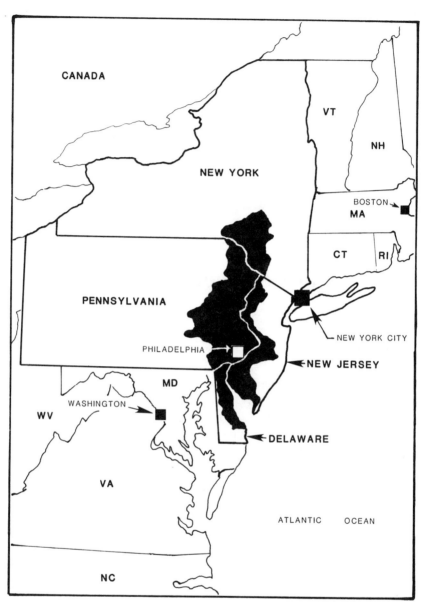

LOCATION OF THE DELAWARE RIVER BASIN

Chronology: Evolution of a Delaware River Dam

1783	New Jersey and Pennsylvania antidam treaty
1823–25	Lackawaxen Dam built (only true dam ever built on the Delaware River)
1842	Start of the New York City water system
1850	First Philadelphia intake on the Delaware
1852	First Supreme Court case involving the Delaware
1858	First Philadelphia look at an upland water supply
1883–85	First Philadelphia study to look at an upland water supply involving use of the Delaware
1900–20	Delaware examined for hydropower
1908	First meeting of New Jersey, New York, and Pennsylvania concerning the Delaware
1920–50	New Jersey, New York City, Pennsylvania, and Philadelphia produce numerous water studies involving the Delaware River
1921	First New Jersey water-supply study to propose use of the Delaware
1924	First study of a water-supply dam at Wallpack Bend
1925	First Tri-State Delaware River Compact
1927	Second Tri-State Delaware River Compact
1928	New York City decides to tap the Delaware

1929   Horton power plan for Trenton is first to consider multiple-purpose water development

1929–31   Delaware River Case argued before U.S. Supreme Court

1930–34   Corps of Engineers produces 308 Plan, the first comprehensive water-resources plan for the Delaware; Tocks Island Dam is first proposed

1936   Interstate Commission on the Delaware River Basin created

1938   New Jersey proposes use of the Delaware and Raritan Canal for water-supply purpose

1942   Test borings at Tocks Island lead to rejection of site; Wallpack Bend now favored

1945   Philadelphia Board of Consulting Engineers proposes a dam at Wallpack Bend *and* use of the river to deliver water to Philadelphia

1948   Delaware River Basin Resources Development Compact is written but not used by Incodel

1951   Multi-dam project proposed by Incodel; Delaware River Basin Water Commission proposed to build Incodel water-supply project

1952–54   Second Delaware River Case argued before U.S. Supreme Court

1955   Pennsylvania begins study of a state-built dam at Wallpack Bend; begins land acquisition; record flood hits the Delaware in August

1956   Corps of Engineers begins comprehensive study of the Delaware River Basin; Delaware River Basin Advisory Committee created

1957   Corps releases results of special Tocks Island study; Tocks Island replaces Wallpack Bend

1959–61   Delaware River Basin Compact is drafted and adopted by states and the federal government

1960   Corps reservoir plan is finalized (Tocks Island is the largest dam in the plan); power companies announce plans for the Kittatinny Mountain Pumped-Storage Project involving Tocks Island Dam

1961   Power companies purchase land from New Jersey, including Sunfish Pond

1961–67   Record drought hits Delaware River region

1962   Tocks Island Dam Project authorized by Congress

1964   Tocks Island Dam receives first appropriations

1965   Delaware Water Gap National Recreation Area established around Tocks Island Reservoir; Delaware Valley Conservation Association formed to fight dam and recreation-area projects

1966   Lenni Lenape League formed to fight pumped-storage project

1970   National Environmental Policy Act becomes law

1971   Corps delays start of Tocks Island Dam because of incomplete environmental impact statement; Save the Delaware Coalition formed to fight dam

1972   Congress delays dam construction until environmental problems are resolved; New Jersey's Governor Cahill outlines conditions for his state's continued support of dam

1974   Congress budgets $1.5 million for Tocks review study; for first time no money is appropriated for dam project; first bill introduced to deauthorize dam project

1975   Delaware River Basin Commission votes 3 to 1 against the construction of Tocks Island Dam

1975–77   Attempts to deauthorize Tocks fail

1976   Delaware River Basin Level B study begins

1978   Middle Delaware Scenic and Recreational River created in Tocks Island region

1979   Good Faith negotiations begin

1980–81   Drought declared; Tocks Island discussed

1983   Good Faith agreement signed by parties to U.S. Supreme Court case; Delaware River Basin Commission adopts the Good Faith recommendations; construction of Tocks Island Dam deferred until sometime after the year 2000

1984–95   Timetable for implementing the Good Faith agreement

1985   Drought declared; ground breaking for the first Good Faith project, the Merrill Creek Reservoir, in September

After 2000 Final decision on Tocks Island Dam?

List of Abbreviations and Acronyms

CEQ   Council on Environmental Quality
cfs   cubic feet per second (a measure of stream flow)
Corps   U.S. Army Corps of Engineers
DRBAC   Delaware River Basin Advisory Committee
DRBC   Delaware River Basin Commission
DVCA   Delaware Valley Conservation Association
DWGNRA   Delaware Water Gap National Recreation Area
EDF   Environmental Defense Fund
Incodel   Interstate Commission on the Delaware River Basin
mgd   million gallons per day (1 mgd = 1.55 cfs)
Minisink   portion of Delaware River Valley upstream of the Delaware Water Gap
TIRAC   Tocks Island Regional Advisory Council
TIRES   Tocks Island Regional Environmental Study
WRA/DRB   Water Resources Association of the Delaware River Basin

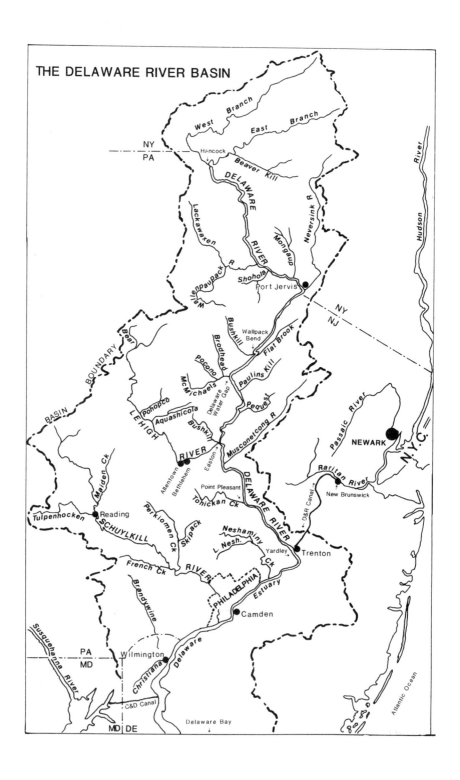

THE DELAWARE RIVER BASIN

1

Introduction

On lists of great American rivers, the Delaware River is often missing. Almost sixty U.S. rivers are longer, and thirty-two of these carry more water than the Delaware.[1] The Delaware's average discharge is less than 3 percent of the Mississippi's and only 46 percent of its western neighbor, the Susquehanna. Its drainage area, 12,765 square miles, encompasses only 0.4 percent of the land area of the United States. No other river system of this size, however, provides almost 10 percent of the U.S. population with water. Surprisingly, the Delaware River is also one of the last major U.S. rivers without a dam on its main stem.

In August 1955 a severe flood ravaged the Delaware River and its tributaries. Property damage was extensive, and one hundred lives were lost. The flood marked a turning point in the water history of the Delaware River Basin. Within seven years the basin would have a massive multi-reservoir plan authorized by Congress and a unique federal-state water compact adopted to implement the plan. The keystone of the reservoir plan was a dam across the Delaware River at Tocks Island. Tocks Island Dam was to be the U.S. Army Corps of Engineers' eighth largest undertaking[2] and its largest dam project east of the Mississippi River. Associated with the dam project was a national recreation area, which was to be the busiest unit in the National Park System, and a large pumped-storage electrical generating station to be owned by three power companies. For some people, the dam project was the type of activity that excited the imagination and glorified the engineer in man.

Aerial view of Tocks Island in the middle of the Delaware River. Perched on top of Kittatinny Ridge is Sunfish Pond. The dam site is marked by the power line crossing of the river. (Leigh Photographic Group, DRBC Collection)

It is often stated that Tocks Island Dam was born in the floodwaters of 1955. This is not so. A dam across the free-flowing Delaware River had long been unfinished business for the basin's water engineers. It was the product of decades of controversy surrounding the use of the river system and a natural reaction to the demands being placed on it. Four states and two large cities had stakes in the Delaware, and one of these was located outside the basin. New York City had fought in the U.S. Supreme Court for the right to export Delaware River Basin water to the Hudson and had won. The entry of New York City into the Delaware's water affairs had profound ramifications and focused the attention of New Jersey, Pennsylvania, and Philadelphia on the need for a Delaware River dam. Whether this need was real or imagined can only be debated.

While Tocks Island Dam was the product of an evolutionary process, it was also part of the process. The dam project promised much and received widespread support in its early years. It was large enough, however, to generate controversy. The project involved massive federal land acquisitions and the dislocation of thousands of people. Moreover, the dam created environmental impacts that were large enough to cause great alarm. Delayed by the Vietnam War and caught up in the environmental fervor of the early 1970s, Tocks Island Dam created a firestorm of national proportions. Inherent in this controversy was the issue that plagues every dam controversy: Was the river to serve mankind, or remain free-flowing?

In 1975 three of the four Delaware River Basin states voted against building Tocks Island Dam. In the aftermath of this decision, the Delaware River in the Tocks region was added to the National Wild and Scenic Rivers System. The demise of the dam project sent shudders through the Delaware River Basin water community. In this vacuum, however, new water-planning activities were initiated. These ultimately led to the adoption of a historic interstate water-management agreement for the Delaware River Basin, one involving no dams on the Delaware River. This plan is now undergoing implementation.

Water management in the Delaware today owes much to the evolutionary process that created Tocks Island Dam and then killed it. An understanding of this process is important not only for today but for the future. Water management in the Delaware continues to evolve, and future decisions concerning the river will be important ones. More than likely, scarce resources like water and free-flowing rivers will be debated.

This book is an attempt to describe and explain the historical context of today's Delaware. By necessity, a major focus of this history is the longstanding dream of a dam across the main stem of the river. Tocks Island Dam was the culmination of that dream, and the histories of the river and the dam are inextricably intertwined.

2

Enter Man

The Delaware River story began about 6 million years ago. To the east of North America was another continent, the forerunner of Europe and Africa. For about 400 million years, the giant land masses collided and pulled against each other. The geology of the Delaware River Basin is the history of this contest. When the continents separated for the last time, the Delaware River was created. At first it was a sluggish river flowing across flat peneplains left after the erosion of the third-generation Appalachian Mountains. Sometime later, today's mountains began rising and the river was rejuvenated. The river kept its ancient course by cutting water gaps as the new mountains rose. Still later, the ice ages carved the final features of today's Delaware River Basin. The geological events that occurred over millions of years gave the region its unique character.

The first people to live in the Delaware River Basin arrived shortly after the last ice age, about 7,000 to 12,000 years ago. They were gone by the time that the Lenni Lenapes (or Delaware Indians) arrived, about 200 years before Columbus discovered America. The Lenni Lenapes settled throughout the region and prospered until the coming of the Europeans.

The first European discovery of the Delaware River system occurred on August 28, 1609, when Henry Hudson and the crew of the Dutch ship *Half Moon* discovered Delaware Bay. The English repeated Hudson's discovery one year later, when the captain of the *Discovery*, Samuel Argall, also discovered Delaware Bay. Argall named what is now Cape Henelopen for Sir Thomas

West, the governor of the Virginia colony. West's royal title was Lord De La Warre, and it eventually was applied to the whole river system.

The Dutch ruled the Delaware River region until 1664, when its territories were captured by the English. Under English rule, the region grew and prospered. Not far from the banks of the Delaware River, however, the thirteen English colonies declared their independence on July 4, 1776. Many important events in the American Revolution occurred in the Delaware River region, and it remained a focal point for the nation after independence was won. It was shortly after the war that the water resources of the region began taking on new importance.

Navigation was the first water concern in the nontidal Delaware River. Around 1770, private contributors raised money to improve passage of coal arks, Durham boats, and lumber rafts traveling through the rapids at Trenton. A channel was cleared through the rocks and the route marked. In the following year, Pennsylvania authorized a private project to improve navigation north from Trenton to the Neversink River. The American Revolution delayed this work until 1791, when private contractors were hired to blast out the larger rock obstructions, particularly at Foul Rift. Like the earlier work, the improvements resulted in a channel through the worst places.

Recognition of the importance of the river traffic went beyond the removal of rocks that were breaking up Durham boats, coal arks, and rafts. In April 1783 New Jersey and Pennsylvania appointed commissioners to decide the ownership of each Delaware River island. The 1783 treaty, however, did much more. It also sought to protect the passage of lumber rafts and other river craft by declaring that the "whole length and breadth (of the Delaware River) thereof, is and shall continue to be and remain a common highway, equally free and open for use, benefit and advantage of each state." This meant that no one could build a dam across the Delaware River unless both states repealed the treaty. The exception was wing dams, low dams that are broken in the middle or that extend only partway into the river. Under a different interpretation, the treaty clause also meant that water could not be diverted from the river if the diversion substantially reduced river flows.

The 1783 treaty grew in importance with the rapid growth of the Delaware River lumber-rafting industry after 1800. Wing dams and water diversions would become thorny issues between the two states for many years. Most of the wing dams served mills or were built to divert water from the river into an adjacent canal. The two states nearly went to the U.S. Supreme Court in 1815 when New Jersey authorized a mill dam without Pennsylvania's permission. During the dispute, the two states conducted a survey of wing dams between Trenton and Belvidere. They found that Pennsylvania had more than New Jersey and that most interfered with the passage of river craft.

After 1800 various technological advances were made, and the Industrial Revolution slowly got under way. The changes wrought by the Industrial Revolution centered on the application of steam power to manufacturing and transportation. The perfect fuel for the Industrial Revolution had been found in the upper Lehigh watershed in 1791 by Philip Ginder. There and in the nearby Schuylkill area were vast underground deposits of anthracite coal, a remnant of the great prehistoric swamps that had once occupied the area.

The coal in the upper reaches of the Delaware River Basin was useless until it could be moved efficiently to market. In 1823, therefore, one of the nation's first canals was opened to the Lehigh coalfields. The first coal shipment down the seventy-two-mile-long Lehigh Canal marked the start of the Industrial Revolution in North America.[1] The canal company, the Lehigh Coal and Navigation Company, would later play a major role in keeping water-supply dams from being built on the Delaware River.

The success of the Lehigh Canal stimulated the construction of other canals: the Morris Canal (Phillipsburg to Newark, New Jersey); the Delaware Division of the Pennsylvania Canal (Easton to Bristol, Pennsylvania); and the Delaware and Raritan Canal (Bulls Island, Bordentown, and Trenton to New Brunswick, New Jersey). These canals linked the Easton terminus of the Lehigh Canal with New York City, Philadelphia, and the cities of New Jersey. The coal generated industrial growth in each city, which, in turn, led to massive increases in population.

Two other major canals linked the coalfields with Philadelphia and New York. The Schuylkill Canal was an important shipper of coal to the Philadelphia area, while the Delaware and Hudson Canal hauled coal to New York City. Both canals contributed to the intense commercial rivalry between the two great cities. The Delaware and Hudson Canal is of particular interest since the canal company built the only true dam ever constructed across the Delaware. This sixteen-foot-high dam, built at Lackawaxen, Pennsylvania, between 1825 and 1828, raised the river level so that canal boats could cross. The river crossing was replaced in 1849 by the now-famous Roebling Aqueduct, but the dam itself lasted into the twentieth century.

The implications of the 1783 bi-state treaty changed with the construction of canals. New Jersey, in particular, disliked the diversion of Lehigh River water into the Delaware Division Canal at Easton. Another New Jersey concern was the wing dam at New Hope, Pennsylvania. Portions of this wing dam may have dated to 1812, but in 1831 the dam was enlarged and a waterwheel constructed to divert river water into the Delaware Division Canal.

The water diversions were not unilateral, however. On the New Jersey side of the river, the original Delaware and Raritan Canal was given a charter by

New Jersey in 1824. Because of the bi-state treaty, the canal company's charter required Pennsylvania's consent. Pennsylvania's requirements for its consent, however, were unacceptable, and the canal was not built.[2] During the controversy, two Pennsylvania mill owners, George Rundle and William Griffiths, took legal actions to keep the proposed canal from diverting Delaware River water. The New Jersey Circuit Court concluded in 1828 that the state did not have to get Pennsylvania's consent to the water diversion as long as navigation was not disrupted. The case ultimately went to the U.S. Supreme Court. Its 1852 decision (*Rundle v. the Delaware and Raritan Company*) concluded that the mill owners could not raise interstate issues unless one of the states did so first.

By the time the Rundle case made it to the U.S. Supreme Court, the Delaware and Raritan Canal had already been built by another company. The canal diverted river water at Bulls Island, opposite Lumberville, Pennsylvania. At the Bulls Island inlet, the Delaware and Raritan Canal Company erected a wing dam to raise the water level at the canal inlet. Later the canal company built another wing dam at Lambertville, New Jersey, immediately opposite the New Hope wing dam of the Delaware Division Canal. The Delaware and Raritan Canal diversion would become a key New Jersey water-supply facility many years later.

The antidam treaty was largely forgotten with the construction of the last canal improvements. Interest in the Delaware River would rise again, however, with the development of a new energy source, electricity. By the 1890s, the uses of electricity were rapidly being extended, from city streets to factories. Its impact on Americans would be as revolutionary as steam power was at the beginning of the Industrial Revolution.

One method of generating electricity is to use the fall of rivers and streams to drive turbines. New Jersey may have been the first state to recognize the potential of the Delaware River for hydroelectric power. In 1894 the state's Geological Survey studied the water-supply and power potential of New Jersey's streams, and in 1897 the state legislature enacted a law that allowed companies to be formed specifically for building hydropower dams. The dams were limited to a height of ten feet and were required to have chutes for passage of lumber rafts and shad, indicating that the legislation was directed toward the Delaware River. Various Delaware River–oriented power companies would be organized under this law.

Possibly the first hydropower project proposed for the Delaware River was the 1902 proposal of the Delaware Water Gap Water Power Company. The company supposedly surveyed the region north of the Delaware Water Gap and thus was possibly the first to examine the Tocks Island Dam site. The company was promoted by Philadelphians of whom nothing is known. It is easy to see this project as a summer diversion of wealthy but bored vacationers at one of the Delaware Water Gap resorts.

In 1907 the New York legislature got into the act when it authorized flood-control and power-development studies by the state's Water Supply Commission. In the Delaware River region of the state, the commission studied a potential power system consisting of a large reservoir near the village of Cannonsville on the West Branch and three low dams across the Delaware River at Narrowsburg, Barryville, and either Mongaup or Millrift. The low dams were to be power dams, with river flows regulated by the larger Cannonsville dam upstream. Although the power system had many attractive features, its major disadvantage was the lack of nearby customers.

Concurrent with the New York State study was an even grander scheme to develop the upper Delaware River Basin for power. Upstream from Port Jervis, forty hydropower dams were contemplated, including a series of Delaware River "lakes," each from five to ten miles long. Potential water storage in the system was estimated to be at least 26 billion cubic feet. The developers of the power scheme were reported as "interested persons" in the 1908 annual report of the New York State Water Supply Commission. The report's authors believed that the many lakes would "convert the region into a beautiful and attractive pleasure resort and greatly increase the land values."[3]

Shortly after the New York State studies, a New Jersey company began promoting a hydropower dam near Belvidere, New Jersey. This proposal was serious enough to reactivate interest in the old 1783 antidam treaty. On October 27, 1910, the Delaware River Improvement Company applied to the Pennsylvania Water Supply Commission for approval of the portion of its project lying in Pennsylvania. The commission, however, was not sure of its jurisdiction and requested an opinion from the state's attorney general. The attorney general responded that the 1783 treaty was still valid: "The Water Supply Commission is not vested with power to abrogate this agreement by approving plans for the construction of a dam in the river Delaware. While this agreement remains in force, the only method by which permission could be legally secured to dam the Delaware River, would be by concurrent legislative action of both states."[4] As a result, no dam was built.

While the Pennsylvania legal decision dissuaded the Delaware River Improvement Company from pursuing its plans, the proposal for a dam near Belvidere was soon taken up by others. On January 14, 1913, the Delaware River Development Company was incorporated in New Jersey to develop the same project. Although the company remained active for at least eight years and owned land on both sides of the river, it never overcame the legal obstacles inherent in the antidam treaty. The treaty not only affected the

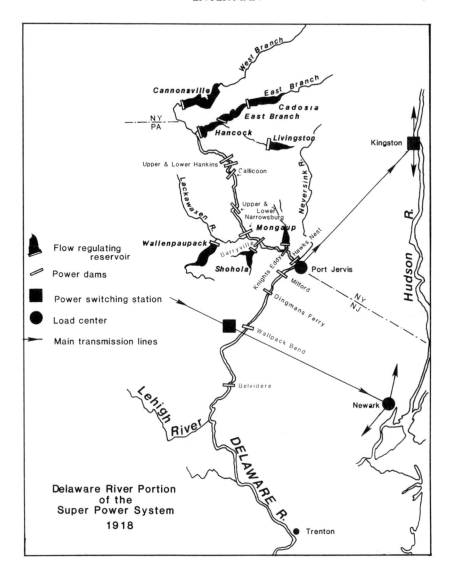

**Delaware River Portion
of the
Super Power System
1918**

Flow regulating
reservoir

Power dams

Power switching station

Load center

Main transmission lines

legal right to build a dam across the Delaware but presumably presented an impediment to any company seeking financing for such a project.

Dams at Belvidere were nothing compared to the "super-power system" proposed in 1918 by E. G. Buckland, the president of the New York, New Haven and Hartford Railroad. The super-power system, stretching from Boston to Washington, D.C., was to be a coordinated system of steam-electric and hydroelectric plants capable of generating a whopping 31 billion kilowatts of electricity. The primary customers for this super-power were to be the eastern railroads, who (it was presumed) would electrify some nineteen thousand miles of tracks. Total cost of the huge system was in excess of one billion (1920!) dollars. Buckland managed to interest Congress in the proposal, and in 1921 the Department of the Interior was appropriated $125,000 to study the idea. This was matched with $26,000 and technical assistance from various railroads and industries.

The super-power project consisted of proposed dams and generating plants throughout the eastern states. In the Delaware River Basin, six flow-regulating reservoirs and sixteen power-generation dams were proposed. The largest regulating reservoirs were planned for Cannonsville, New York, on the West Branch; several on the East Branch; and one on Wallenpaupack Creek. Twelve power dams were planned on the Delaware between Hankins, New York, and Belvidere, New Jersey. Power dams were also proposed for Wallenpaupack Creek, Shohola Creek, and the Mongaup River. The ultimate yield of the Delaware River portion of the power system was believed to be 1.49 billion kilowatt-hours per year, assuming someone was willing to invest millions of dollars. The super-power proposal went nowhere, although it was not from thinking too small.

The early 1920s saw several long-term trends beginning to cross each other in the Delaware River Basin. In 1923 the last of the lumber rafts made its lonely trip down the Delaware River. The lumber-rafting industry had largely been responsible for the Pennsylvania/New Jersey antidam treaty in the first place. By 1920 this treaty was clearly seen as an impediment to the development of the Delaware River for hydropower. No one believed, however, that the antiquated treaty would be a major impediment for long. Interest in developing the Delaware was there—the unused resource represented by the river almost demanded to be developed and exploited.

Interest was rising quickly, however, in another use for the Delaware River. The clean, rushing waters of the upper Delaware River Basin had attracted the attention of the large, thirsty cities to the east and south. Their water-supply engineers were now looking for new water supplies. The Delaware River's future would be largely concerned with water-supply issues and not with hydropower.

3

A Water Supply for Millions

Most American cities were slow to provide adequate, safe water to their citizens. Before the first municipal water systems were even being considered, many cities found themselves faced with exploding populations, annual epidemics of waterborne diseases, and destructive fires. In the 1700s, thousands died annually in Philadelphia and New York City from typhoid, cholera, and other deadly diseases. Fires could be equally deadly. For example, an 1835 New York City fire destroyed seven hundred buildings and bankrupted every insurance company operating in the city.[1] The epidemics and fires, accompanied by the failure of private enterprise to cope with the problem, led after 1800 to the first public water systems.

The beginnings of the Philadelphia water system can be traced to the death of Benjamin Franklin in 1790. Franklin left the city funds for a public water-supply system. With these and other funds, Philadelphia opened its first waterworks on January 27, 1801. This system, with intakes on the Schuylkill River, expanded over the decades. By 1850 the Philadelphia system was supplying seven million gallons of water to its residents. In that year the suburb of Kensington (now part of Philadelphia) built its own waterworks. The construction of the Kensington works on the Delaware River marks the beginning of Philadelphia's use of the nearby Delaware Estuary for water. Ultimately, the city would expand this portion of its system until the Delaware was providing fully 50 percent of the city's water.

The Delaware and Schuylkill rivers would give Philadelphia enough water to accommodate a growing population and industrial base. The quality of the city's water, however, was poor, and it would get much worse as the city grew and sewers were built. These dumped tremendous amounts of wastes into the nearest watercourses: the Schuylkill and Delaware rivers. Upstream from Philadelphia, many other cities, towns, and industries did the same thing. These wastes also found their way to Philadelphia. The result was obvious. Philadelphia's ample, nearby water supplies were soon grossly polluted.

New York City's water system began much later than Philadelphia's. The disastrous 1835 fire was the impetus that finally forced the city to build a municipal water-supply system. After the fire, funds were authorized to build a reservoir in the Croton watershed in Westchester County. A forty-two-mile-long aqueduct, receiving reservoirs, and other parts of a water-distribution system were built. In the summer of 1842, the first water was delivered to new receiving reservoirs in today's Central Park and at 42nd Street. At the time, New York's water system was one of the most advanced in the country.

The capacity of the original Croton system was exceeded six years after it was completed. Between 1842 and 1911, the Croton system was expanded almost continuously until it was yielding 104 million gallons per day. Triggering the various expansions were water shortages in 1850, 1864, 1865, 1869, 1870, 1876, and 1877, and less serious concerns in other years. As the water shortages indicate, New York City's problem was how to keep pace with an ever-increasing demand for water.

New York City's water problem was two-dimensional. Rapid population growth was one dimension. Between 1850 and 1900, the city grew from a half-million persons to more than three-and-a-half million. By the turn of the century, more than 135,000 people were moving into New York City each year. The rapid population growth put incredible demands on the city's ability to provide water.

The second dimension of the demand problem was increasing consumption on a per capita basis. The customers of the first water system had felt content to use about 20 gallons of water per day. By 1895, however, New Yorkers were consuming an average of 145 gallons per person per day. Of course, each individual was not using this much water. Industries and commercial establishments were contributing to the high average.

Philadelphia and New York City had taken distinctly different approaches to solving their early water problems. Philadelphia, the city of conservative Quakers, had taken the least costly approach by tapping ample, nearby water sources. However, its water supplies had become grossly polluted. New York City, the more flamboyant city, had taken the bold approach of building reservoirs in distant areas and then bringing the water to its citizens. New York City's approach provided clean water, but it had a flaw, too. The city's construction program could not keep pace with the tremendous growth in population and water consumption.

Philadelphia eventually began exploring the possibility of developing its own upland sources of water. In 1858 Henry P. Birkinbine, the city's chief engineer, proposed damming streams twenty-five to thirty miles north of Philadelphia and piping the water to the city. Shortly after the Civil War, Birkinbine developed a specific plan consisting of a reservoir on Perkiomen Creek in neighboring Montgomery County and a gravity-delivery system to bring the water to Philadelphia.

Birkinbine's proposal was examined in 1875 by the Philadelphia Water Commission, but nothing was done. Pollution continued to worsen, however, leading to a series of technical studies in 1883–85 by Rudolph Hering. Two alternatives emerged from these studies. One was an expanded and modified version of the Birkinbine proposal. Under this alternative, a reservoir on Perkiomen Creek was to be constructed as the first step in the development of a much larger system that ultimately saw reservoirs built in the upper Lehigh watershed.

Hering's second alternative called for a reservoir on Tohickon Creek in Bucks County, Pennsylvania. At Point Pleasant, a water-powered pump was to be built on the Delaware River in order to pump river water to the reservoir. This water was then to be delivered to Philadelphia. The Delaware–Tohickon Creek system was also considered to be the first step in the construction of a future mountain water-supply system for Philadelphia. When population growth dictated, a conduit was to be built from Point Pleasant to the Delaware Water Gap area. There the city would dam various Pocono streams.

Although Philadelphia's engineering staff recommended that the Point Pleasant alternative be built, nothing was done. A major reason for Philadelphia's inaction was a belief among some influential citizens that Philadelphia should build water-filtration plants instead of expensive reservoirs. They believed that these could render Philadelphia's water fit to drink by filtering out pollutants.

Reflected in the 1885 debates were three conflicting philosophies about the future source of Philadelphia's water. Most of Philadelphia's water debates would consider the same three alternatives: the continued use of the city's nearby but polluted sources of water; the construction of an upland water-supply system in the Lehigh watershed; or the construction of an upland water-supply system in the Delaware Water Gap region. A partial answer to this debate came between 1899 and 1911 when the city built slow-sand filtration plants to treat its water. An upland water system was not considered at this time because the city had severe restrictions on its bonding capabilities. While the filters, the largest of their kind in the world, removed the threat of epidemics from Philadelphia's water, they did little to help the taste and odor. Thus, Philadelphia's dream of an upland water-supply system remained very much alive.

At the same time that Philadelphia was coming to grips with its problems, New York City was grappling with its own. The combination of increasing demand for water and an exploding population exceeded New York City's ability to keep up. Compounding the city's problems was the consolidation of Manhattan with the other New York boroughs. This placed additional stress on the city's abilities since it inherited woefully inadequate water systems in some of the other boroughs. Around 1895 a policy decision was made to solve the city's water problems by continuously expanding the system instead of attempting to decrease water consumption.[2]

New York had one other problem, though. The completion of the large New Croton Dam in 1906 had resulted in the passage of the Smith-Dutchess County Act, which forbade the city from building any more reservoirs in the neighboring counties of New York. As a result, New York City's Catskill System was built between 1907 and 1928. On the eastern side of the Catskills, in the area draining to the Hudson River, the city built Schoharie and Ashokan reservoirs. Connecting these large reservoirs with the city were 144 miles of aqueducts and various other facilities. When the ambitious Catskill project was completed, New York City's water-supply system was capable of providing 879 million gallons of water per day. Unfortunately, the city had not won the race against increasing demand. Before it was completed, it was clear that the Catskill System would be inadequate.

On December 31, 1920, the chief engineer of the city's Board of Water Supply, J. Waldo Smith, reported that water use in New York City would outstrip available supplies by the mid-1930s. The city had three alternatives. The first, more reservoirs in the watersheds east of the Hudson River, was fraught with the political problems manifested in the Smith-Dutchess Act. The second alternative was to tap the Hudson River somewhere upstream of the city. This alternative was not favored, however, because the pollution in the lower river would have required the city to build water-treatment plants. This left the third alternative: development of the watersheds on the Delaware River side of the Catskill Mountains. From New York City's perspective, it was clearly time to talk to New Jersey and Pennsylvania about their plans for the Delaware.

New York City was not the only city in 1920 that was beginning to take a serious look at the Delaware. Philadelphia had not considered the possibility of an upland water-supply system since the debates of 1885. The taste of Philadelphia's water was now the butt of vaudevillian jokes. In 1920, therefore, Philadelphia's mayor appointed a panel of experts to look for a new source for the city's water. The panel, however, concluded that a mountain water supply was too costly. Instead, they recommended the construction of reservoirs in the Perkiomen, Tohickon, and Neshaminy watersheds in nearby Bucks and Montgomery counties.

The recommendations of the mayor's panel were controversial. The rejection of a water supply in the mountains angered many Philadelphians who wanted an upland source of clean mountain water such as that enjoyed by New York City. Another panel was therefore appointed in 1924. This panel recommended that Philadelphia expand and upgrade its filtration plant on the Delaware Estuary and abandon the Schuylkill River. The Schuylkill water supply was to be replaced by reservoirs in the Perkiomen and Tohickon watersheds. The panel rejected a so-called Blue Mountain water-supply system that consisted of four reservoirs in the upper Lehigh, five on Delaware River tributaries in the Pocono Mountains, and connecting conduits.

The Blue Mountain system was not rejected outright. For one thing, the panel did not recommend developing the Neshaminy watershed because it involved an intake on the Delaware River at Yardley, Pennsylvania. This was not considered a desirable location for an intake—if a Blue Mountain project was to be built at a later date. Furthermore, the panel predicted that the city would eventually get its water from the mountains. They believed that the Blue Mountain system was "unquestionably the best of all (alternatives) which have been considered for the City of Philadelphia, if we leave but the question of costs."[3]

While Philadelphia was studying its alternatives, the Pennsylvania Water and Power Resources Board was conducting a study of its own. This study would come up with a more revolutionary alternative for Philadelphia. From its data and projections, the board's staff concluded that a dam across the Delaware River at Wallpack Bend would be the answer to Philadelphia's needs. Such a dam could provide 750 mgd of water to the city, using a fifty-eight-mile-long system of conduits, tunnels, and inverted siphons. The cost of the plan was estimated to be $70 million, which was much less than any of the alternatives that Philadelphia's panel of experts had examined. The Pennsylvania Water and Power Resources Board study was an important milestone since it was the first to propose a Delaware River water-supply dam.

Philadelphia and New York City were not alone in their search for water in the early 1920s. New Jersey also had its water needs, particularly in the densely populated North Jersey area. More than two million people lived in North Jersey, and the area had an annual growth rate greater than that of New York City. Beginning in 1921 New Jersey and its North Jersey Water Supply District would begin a series of water studies for North Jersey.

The first, the Hazen study, done for the New Jersey Department of Conservation and Development, came up with four potential water-supply projects for North Jersey. Two of the alternatives led ultimately to the Delaware River. The first of these involved the construction of a reservoir on the Raritan River above Somerville, New Jersey. While this reservoir was large enough to satisfy North Jersey's needs for some time, it was proposed that the reservoir be augmented by pumping Delaware River water into the Raritan river near Clinton, New Jersey, when additional growth required more water. A pump station was planned on the Delaware near Belvidere.

The favored alternative was the so-called Long Hill project, which called for eight interconnected reservoirs to be built over many decades. The Long Hill reservoir near Morristown was to be built first. As growth occurred, reservoirs were to be built on various Delaware River tributaries. The final stage in the project was the construction of a pump station on the Delaware River near Wallpack Bend. The pump station was to pump Delaware River water to a nearby reservoir on Flat Brook. The entire system would provide at least 1,500 mgd, or twice the amount of water North Jersey would need by 1970. Nothing was built as a result of the study, however.

The Long Hill system was restudied by New Jersey's North Jersey Water Supply District in 1925. Instead of the Long Hill reservoir, the study recommended the construction of Chimney Rock reservoir near Bound Brook. The Chimney Rock proposal also expanded west to New Jersey's Delaware River tributaries. The final stage in this system was to be the construction of a pump station on the Delaware River near Belvidere.

Although the cities of Newark and Elizabeth verbally agreed to build the Chimney Rock project in 1927, it was never built. Alternatives to the Long Hill and Chimney Rock systems would be studied at different times, but none of these was built either. New Jersey, like Philadelphia, had a hard time making up its mind. One thing appeared sure by the 1920s, however. The ultimate solution to the water needs of North Jersey seemed to lead to the Delaware River.

By the early 1920s, it was clear that New Jersey, New York, and Pennsylvania needed to get together to discuss the Delaware. The three states had gotten together once before, after New York State had proposed its large power project for the upper Delaware. In the spring of 1908, representatives of the three states had met in New York City to discuss in general terms the development of the Delaware River Basin and other shared watersheds.[4] The discussions ended with a call for cooperation among the three states and the federal government in the Delaware River Basin.

As it turned out, the 1908 discussions were premature. The situation had changed by the 1920s with the new interest in using the Delaware for water supplies. Meetings were held, and in 1923 the Delaware River Treaty Commission was created by legislative action in the three states. The commission's job was to negotiate an interstate agreement that would govern each state's water projects in the Delaware River Basin. Federal interest in the Delaware was recognized, and federal agencies were invited to participate as advisors. Delaware, the fourth Delaware River Basin state, had no rights to the water of the upper Delaware and was not asked to participate. It had not yet realized its own stake in the discussions.

The actual work of the commission was performed by staff of the various state agencies, New York City, the Federal Power Commission, and the U.S. Army Corps of Engineers. Detailed flow studies were completed in November 1924, and by mid-December a draft interstate compact had been developed. The interstate compact was a new, almost radical, approach for allocating water among states. The Colorado River Compact, which had been ratified two years earlier, was the nation's first. On January 24, 1925, the eight commissioners of the Tri-State Delaware River Commission (as it was now called) fixed their signatures to the proposed compact and recommended that each of the states adopt it.

Under the proposed compact, Pennsylvania, New Jersey, and New York shared equally in the waters of the Delaware River Basin above Port Jervis. Below Port Jervis, the Delaware was shared equally between Pennsylvania and New Jersey. Each state was allowed to develop three-fifths of its drainage area, provided that certain stream flows were maintained during the dry period from July to October. The intent of the compact was to maintain a flow in the Delaware River of 0.45 cfs per square mile of drainage area above Port Jervis and 0.36 cfs per square mile of drainage area below Port Jervis. (These flows corresponded to a flow of 1,400 cfs at Port Jervis and 2,700 cfs at Trenton.)

The tri-state compact was even more comprehensive. A permanent commission, the Tri-State Delaware River Commission, was to be established. The commission was to conduct comprehensive water-resources planning, collect stream data, and perform various regulatory functions. In addition to water supply, the new commission was to deal with stream pollution and other aspects of water-resources development.

The 1925 compact for the Delaware River was clearly ahead of its time. The participants in the compact process were genuinely pleased with their efforts. Soon, however, various criticisms and fears concerning the compact began to be heard. Parochial and special interests were behind some of these criticisms and fears; others were politically motivated. Even the New York Times, which was very much in favor of the compact, cautioned in a January 29, 1925, editorial that the New York legislature should take nothing for granted.

After the compact was signed, it was submitted to each state for passage. New Jersey immediately suggested several interpretive amendments. The treaty, with New Jersey's suggested amendments, passed the Pennsylvania Senate on February 10, 1925, and went to the Pennsylvania House. Passage of the compact in 1925 by Pennsylvania was critical since the state's legislature met only every other year. New York State's legislature passed the compact bill (as amended) without delay, and the governor signed the legislation on March 18. New York State (i.e., New York City) was furthest along in its plans for the Delaware and thus had the most to gain from the compact's immediate passage.

Even though the Pennsylvania Senate and New York State had agreed to its amendments, the New Jersey legislature balked at passing the bill. Instead, the New Jersey Chamber of Commerce was designated as the state's Water Policy Commission and charged with examining the compact. Meanwhile, it became clear that Pennsylvania would not pass the treaty bill either. Some people were bothered that New York City would be the immediate beneficiary of the compact, even though it was outside the Delaware River Basin. Others pointed out that Pennsylvania had a larger percentage of land in the Delaware River Basin than the other two states and thus should be given proportionately more water rights.

Among those lobbying against the bill's passage in Pennsylvania was the Lehigh Coal and Navigation Company. By virtue of its wealth, this company had many friends in Pennsylvania politics. Moreover, the company had been granted water rights to the Lehigh River during the building of the Lehigh Canal one hundred years earlier. Philadelphia and its surrounding counties were logical future customers for these rights. Thus, the Lehigh Coal and Navigation Company had a personal interest in seeing that the Delaware River was not developed for water supply. Because of the opposition, the Pennsylvania legislature concluded its 1925 session without taking action on the compact legislation. It would not meet again until 1927.

On January 30, 1926, the New Jersey Water Policy Commission released its findings. The commission recommended that the proposed compact be rejected because of the different ways some sections could be interpreted. The criticisms made by the commission, however, were meant to be constructive. The commission called for the start of new compact negotiations, not only for the Delaware River, but also for several smaller rivers shared with New York State.

The New Jersey veto of the 1925 compact sent the three states back to the negotiating table. The negotiations were conducted quickly since much of the background work had already been done. On January 13, 1927, the compact commission met in New York City and signed a new "Compact as to the Waters of the Delaware River." The new compact was less ambitious. Instead of trying to divide all the waters of the Delaware River system, the compact gave each state an initial allocation of 600 million gallons per day. Pennsylvania, however, received an additional 300 mgd since it was assumed that Philadelphia's intake at Torresdale would be moved upstream to Yardley, the lower boundary of the treaty. With the extra allocation, the commission also hoped to counter arguments that Pennsylvania was getting less water than it deserved.

The in-stream flow maintenance requirements of the 1927 compact were identical to the previous compact, with one important difference. In the lower portion of the Delaware River, the flow criterion was reduced from 0.36 cfs per square mile to 0.24 cfs. In addition, New York State was given no vote on diversion matters occurring below Port Jervis. This change was made to counter charges that the compact jeopardized the states' rights of the downstream states. Most other provisions of the 1925 compact, however, remained the same.

The drafters of the new compact hoped that the changes would overcome the opposition generated by the first compact. New York again acted quickly, and the compact treaty was signed into law on April 7, 1927. The bill, however, stalled in Pennsylvania, with both proponents and opponents content to take a "wait until New Jersey does" approach. The City of Philadelphia did not agree with this and passed a resolution endorsing the compact on March 31, 1927.

It was in New Jersey that the compact was again debated to its fullest. Both the Senate Judiciary Committee and the Water Policy Commission held public hearings on the compact in the spring of 1927. The focus of the debate was the reduced flow-maintenance requirement in the river below Port Jervis. Many feared that the Delaware would become a "mere brook" during the summer low-flow season.[5]

Leading the opposition to the compact was the City of Trenton, which was concerned that the compact would reduce the amount of water coming down the Delaware. At issue was not the amount of water per se but its impact on quality. The city was already experiencing problems during low-flow periods when thunderstorms washed coal and industrial wastes out of the Lehigh River Valley. This pollution forced the city to shut down its water-filtration plant for a couple of days. Under the low-flows inherent in the 1927 compact, Trenton feared that it would be forced to shut down its water system for weeks at a time or abandon the Delaware altogether.

Trenton's arguments concerned the New Jersey legislature because the analyses of Trenton's engineers differed greatly from those of the compact commission's engineers. A special committee was sent by the legislature to

Pennsylvania's legislature to see if Pennsylvania would agree to an increase in the minimum flow requirements. Nothing happened, however, and New Jersey's internal debate continued into 1928. By extending the debate into 1928, New Jersey had killed the compact. The Pennsylvania legislature did not meet in 1928, a fact known by the now impatient City of New York.

New York City's interest in Delaware River Basin water had been the main factor leading to the compact negotiations in 1923. While the negotiations were occurring, the city's Board of Water Supply had been studying various alternatives for increasing the city's water supply. On October 9, 1926, the board recommended that new water supplies be secured *outside* the Delaware River Basin because of the uncertainty surrounding the passage of the 1925 compact. In spite of this recommendation, New York City's studies had shown that it was cheaper to develop its next water supply in the Delaware.

With the drafting of another compact in 1927, New York City renewed its interest in building dams in the Delaware River Basin. On July 27, 1927, the Board of Water Supply withdrew its earlier recommendations and proposed instead a three-phase plan for developing water supplies in New York State's portion of the Delaware River Basin. The project was designed to conform to the requirements of the 1927 compact, with New York State's entire alloca- tion, 600 mgd, being diverted to New York City. A large reservoir was planned for the East Branch, with smaller ones in the West Branch, East Branch, and Neversink watersheds.

At the suggestion of New York City's mayor, Charles E. Hughes and John W. Davis were hired to analyze the city's legal rights in the Delaware. Hughes and Davis were top legal talent: Davis was a well-known attorney, and Hughes was an attorney and former member of the U.S. Supreme Court. Both had once run for president. At the same time, New York City officially notified Pennsylvania and New Jersey that it intended to build dams in the Delaware River Basin. In Albany the city submitted its plan to the State Water Power and Control Commission and had legislation introduced giving the Board of Water Supply permission to operate in the upper Delaware area. Conflict with Pennsylvania and New Jersey was inevitable.

On April 17, 1929, Hughes and Davis submitted their opinions. Both agreed on the essential point: There was no reason why New York should not have a share of the waters of the Delaware. This principle had been recog- nized by both Pennsylvania and New Jersey during the two compact negotia- tions. Thus, these states would have to prove that the compact materials were incorrect and that injury could not be averted either through operating restric- tions or flow-compensation requirements. Following the release of the legal opinions, the State Water Power and Control Commission gave its blessing to New York City's plans in the Delaware River Basin.

New Jersey's reaction to the legal opinions was simple enough. The state attorney general's office began preparing a lawsuit for the U.S. Supreme Court. On May 4 the New Jersey legal team requested permission from the Court to sue New York State and New York City. This was granted on May 20, 1929. Both New York State and New York City had agreed to the suit in order to expedite the whole process. New Jersey's suit sought to keep the city from diverting any water out of the Delaware River Basin.

After much internal deliberation, the Commonwealth of Pennsylvania entered the case on February 10, 1930, as an intervener. The City of Philadelphia also tried to join the suit, but this was rejected by the Court. A special master, Charles N. Burch of Memphis, Tennessee, had been appointed by the Court on January 27, 1930. His job was to take testimony from all parties, make findings of facts and conclusions of law, and propose a decree for the Court. Five sets of hearings were held, and more than 150 attorneys, experts, and lay witnesses were heard. The testimony and supporting exhibits eventually filled sixty volumes containing ten thousand printed pages.

New Jersey's complaint against New York City was based on four main arguments:

1. That the diversion would injure New Jersey's water-power potential, river recreation, potential water supplies, shad fisheries, industrial water supplies in the Delaware Estuary, oyster harvests, sanitation, and the cultivation of river bottomlands;
2. That the nontidal Delaware River was a navigable river for its entire reach from Hancock, New York, to Trenton; thus, Congress or the Secretary of War had to approve any diversions;
3. That New York City's present use of water was extravagant and that the city had other sources it could develop; and,
4. That New York City's intended diversion violated the doctrine of riparian rights.

New Jersey's contention that New York City's diversion violated the doctrine of riparian rights was particularly interesting since it seemingly sought to prevent any water from being diverted out of the Delaware River Basin.[6] This narrow position was obviously a red herring since adoption of such a policy would have prevented New Jersey from diverting Delaware River water to North Jersey.

Another red herring was New Jersey's power proposal, a system of dams and reservoirs with the ability to generate 350,000 kilowatts of electricity. Designed to operate as a unit, the system consisted of six reservoirs on New York State tributaries (including Cannonsville on the West Branch and near the present-day Pepacton Dam site on the East Branch); a reservoir on the interstate portion of the West Branch; two reservoirs on Pennsylvania tributaries; and twenty reservoirs on the main stem of the Delaware River between Hancock, New York, and Trenton. By proposing locks in the smaller power dams below Easton, New Jersey also claimed harm to future navigation benefits. New York countered that New Jersey's proposed power-development plan was a paper plan for argument's sake. It clearly was.

New Jersey's power proposal was clearly derived from one proposed by Robert E. Horton. Horton first proposed this plan in a 1929 report prepared for the Trenton Public Works Department.[7] Although limited to a "Trenton" perspective, the Horton proposal was in fact the first plan to address multipurpose development of the Delaware River Basin. The plan proposed a minimum regulated flow at Trenton of 6,623 cfs, which was obtained (and presumably financed) by a phased development of the entire drainage area above Trenton. The estimated cost of "Trenton's" development of the Delaware River Basin was about $145 million, with benefits being obtained from hydropower, pollution control (through dilution), navigation (to Port Jervis), and recreation (riverfront parks and cottage developments).

Pennsylvania did not contest New York's right to build dams and divert water out of the basin. It was seeking a Court-approved water allocation of 750 mgd, which was exactly what Pennsylvania thought it could get from a dam at Wallpack Bend. The state also wanted a Court-appointed "river master" to supervise the operations of New York City in the Delaware River Basin.

On May 4, 1931, Justice Oliver Wendell Holmes delivered the opinion of the U.S. Supreme Court. The Court affirmed New York's right to divert water from the Delaware based on the principle of equitable apportionment. According to this principle, each state in the Delaware River Basin had a right to a fair share of the water.

Expounding on the principle of equitable apportionment, Justice Holmes declared that:

> A river is more than an amenity, it is a treasure. It offers a necessity of life that must be rationed among those who have power over it. New York has the physical power to cut off all the water within its jurisdiction. But clearly the exercise of such power to the destruction of the interests of lower states could not be tolerated. And on the other hand, equally little could New Jersey be permitted to require New York to give up its power altogether in order that the River might come down undiminished. Both States have real and substantial interests in the River that must be reconciled as best they may be. The

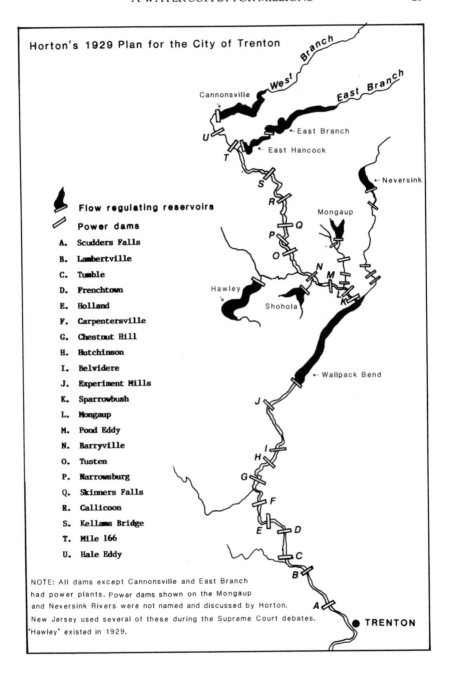

Horton's 1929 Plan for the City of Trenton

Cannonsville

West Branch

East Branch

←East Branch

←East Hancock

←Neversink

Flow regulating reservoirs

Power dams

A. Scudders Falls
B. Lambertville
C. Tumble
D. Frenchtown
E. Holland
F. Carpentersville
G. Chestnut Hill
H. Hutchinson
I. Belvidere
J. Experiment Mills
K. Sparrowbush
L. Mongaup
M. Pond Eddy
N. Barryville
O. Tusten
P. Narrowsburg
Q. Skinners Falls
R. Callicoon
S. Kellams Bridge
T. Mile 166
U. Hale Eddy

Mongaup

Hawley

Shohola

←Wallpack Bend

NOTE: All dams except Cannonsville and East Branch
had power plants. Power dams shown on the Mongaup
and Neversink Rivers were not named and discussed by Horton.
New Jersey used several of these during the Supreme Court debates.
'Hawley' existed in 1929.

● TRENTON

effort always is to secure an equitable apportionment without quibbling over formulas.

The decision of the U.S. Supreme Court in the Delaware River Case outlined the rules that would govern the Delaware for many decades. Specific provisions included:

1. New York City was allowed to divert the equivalent of 440 mgd from the Delaware River Basin. Although it had asked for 600 mgd, the 440 mgd figure satisfied the city since it represented the first two phases of the city's plans for the upper Delaware. New York's diversion was contingent on the city's releasing water from its reservoirs when the flow in the Delaware fell below 0.50 cfs per square mile of drainage area. This flow figure was equivalent to 1,535 cfs at Port Jervis and 3,400 cfs at Trenton.
2. New York City's diversion of 440 mgd was not to be considered a prior apportionment. This meant that sometime in the future the Court could take some of the city's allocation and give it to the other two states.
3. Pennsylvania's requested allocation of 750 mgd was denied. The allocation was denied without prejudice, meaning that Pennsylvania could seek one later. This ruling told Pennsylvania and New Jersey that they would have to have a specific water project in the works (such as that of New York City) if they were to get a share of the Delaware.
4. New York City was required to correct the sewage problem emanating from Port Jervis, New York. Not only was this problem causing water quality problems in Pennsylvania's and New Jersey's portions of the Delaware River, but it promised also to affect water-quality conditions in a future Wallpack Bend reservoir. By 1931 it was generally assumed that a dam at Wallpack Bend would eventually be built.
5. The Court maintained jurisdiction in the Delaware River Case, and each state had the right to apply to the Court for modifications to the 1931 decree. With this provision the Court differed from the recommendations of the special master. Burch believed that the Court's future involvement in the Delaware should be limited to flow-release matters. Pennsylvania's request for a river master to monitor compliance with the decree provisions was also denied. The U.S. Supreme Court's continuing jurisdiction in the water affairs of the Delaware River Basin complicated matters.

The Supreme Court ruling in the Delaware River Case ended a very important decade for the Delaware River. The 1920s had seen water supply emerge as the overriding water-resource concern in the basin, and in 1924 the first water-supply dam had been proposed for the Delaware River. The interest in water supply resulted in the negotiation of two revolutionary interstate compacts. Neither was adopted, however. As a result, an expensive two-year water fight was waged before the U.S. Supreme Court. The rules for playing the water game in the Delaware had now been established.

4

Birth of a Main-Stem Dam Project

The conclusion of the Supreme Court case had no immediate impact on the Delaware River. Although New York City had permission to tap the Delaware, it would be some time before its reservoirs were built. Because of money problems, construction of the Delaware System would not begin until 1937. World War II would cause further delays. The only water projects completed in the Delaware River Basin by 1931 were the construction of the first phase of the Mongaup power project in 1923 and the Lake Wallenpaupack power project in 1926.

The federal government would conduct the next major study of the Delaware River Basin. In 1925 Congress had directed the Secretary of War to prepare a report listing potential rivers and streams where power development might be feasible in conjunction with navigation, flood control, and irrigation needs.[1] The report, with its list of potential study streams, was subsequently published in House Document 308 of the 69th Congress in April 1926. In the following years, funds were authorized for the preparation of various "308 reports," including eight for the Delaware River Basin. The "308" studies were the first comprehensive water-resource studies in the nation.[2] The job of preparing the reports was given to the U.S. Army Corps of Engineers.

The Corps of Engineers had been created in 1775 by the Continental Congress. It was disbanded after the American Revolution but was resurrected in 1802. Early nonmilitary jobs of the Corps were largely limited to internal improvements such as navigation, canals, and roads, but with time the responsibilities of the agency grew. The first all-Corps rivers and harbors bill was passed in 1852, and flood control was added to the Corps' responsibilities in a limited way in 1917. Federal jurisdiction over navigable streams and related tributaries was established by U.S. Supreme Court decisions in 1824 and 1899 respectively.

The Corps' earliest activities in the Delaware River Basin dealt mainly with navigation matters, fortifications, and harbors in the tidal reaches of the Delaware. The first Corps activity in the Delaware River above Trenton was an 1872 investigation that examined the costs of removing navigation hazards (primarily for lumber rafting) between Trenton and Port Jervis. No work was carried out as a result of this study. Similarly, an 1884 study had examined a $7.5 million channelization project between Trenton and Easton, but this project was not carried out either. The same project had been reexamined in 1923–24 with the same outcome. The Corps had also studied ice jams in Port Jervis in 1910 and done other odd jobs through the years.

The preparation of the "308" reports for the Delaware River and seven tributaries was assigned to the Philadelphia District. The Delaware River 308 report was the last of the eight plans to be prepared, and the most important. The study was conducted from 1930 to 1932, reviewed internally for two years, and finally submitted to Congress on January 3, 1934. The Delaware River 308 report was the first comprehensive water-resources plan ever developed for the Delaware River Basin. It not only examined navigation, hydroelectric power, flood control, and irrigation, but also water supply and water quality. Although water-supply planning was not mandated by Congress, the Corps recognized that it was "the most important consideration in connection with any large-scale development of the river."[3]

Navigation, flood control, and irrigation were treated as minor concerns by the 308 report. Various projects were examined, but none was feasible or even warranted. The Delaware River 308 report primarily dealt with potential dams for hydropower and water supply. Thirty-two potential dam sites were evaluated, including dams on the Delaware River at Hankins, Callicoon, Cochecton, Narrowsburg, Barryville, and near the mouth of the Mongaup River, at Wallpack Bend, Tocks Island, Belvidere, and Chestnut Hill (above Easton). Tocks Island was a dam site that no one had looked at in the past. It showed promise.

The Corps' water-supply planning assumed that New York City would be developing a second diversion by 1950. To meet this need the Corps proposed a large reservoir in the Neversink Valley and a Delaware River reservoir near Barryville, New York. New Jersey and Pennsylvania were not ignored. Their water needs were to be obtained from the large reservoir at Tocks Island. New

Jersey's water was to be delivered to a regulating reservoir near Morristown, New Jersey, while Pennsylvania's water would go to Philadelphia via a regulating reservoir near Warrington, Pennsylvania. Making the water-supply projects even more attractive were apparent benefits for low-flow augmentation, pollution dilution, salinity repulsion, and, of course, power production.

Of the thirty-two reservoirs examined by the 308 report, the one at Tocks Island was by far the largest, with 214 billion gallons of planned storage. A reservoir at Wallpack Bend was considered less desirable than one at Tocks Island because it yielded less water. In either case, a dam in the Minisink Valley was considered ideal because the valley was not "occupied by railways or extensive and highly valuable improvements."[4]

Although the Corps found development of the Delaware River Basin to be economically feasible, none of the dam projects justified federal participation. At this time federal participation in water projects was still very limited. Because water supply was believed to be the major use of Delaware River water, the 308 report recommended that an interstate agency be formed "to assure the use of the resource to the advantage of the interested States and cities, and to prevent ill-advised and partial projects that might interfere with major water-supply projects."[5]

Reaction to the Corps' 308 report was minor. Money for large water projects was just not available. In Pennsylvania the hard times created by the Great Depression had brought about the virtual collapse of the state's coal industry. No one was going to suggest building hydropower projects that competed with steam plants burning Pennsylvania coal. The Corps' 308 report for the Tennessee River Valley generated more concern in the Delaware River Basin than the report for the Delaware. When no one stepped forward to build the high dams recommended in the 1930 Tennessee River report, Congress created the Tennessee Valley Authority, a purely federal agency with basinwide authority. In the Delaware River Basin and elsewhere, this piece of New Deal legislation was viewed with great fear by proponents of states' rights.

The Delaware River 308 report did generate two responses. After the report's completion, Philadelphia went through the motion of requesting a $30 million loan from the Federal Public Works Administration for beginning dam construction. Nothing came from Philadelphia's request.

The second response to the 308 report was the incorporation of the Electric Power Company of New Jersey in May 1933. On February 16, 1935, the company applied to the Federal Power Commission for a Preliminary Permit to build hydroelectric dams at Tocks Island, Belvidere, and Chestnut Hill. A project generating 176,400 kilowatts of electricity was proposed at a cost of $29.6 million. Opposing the application were New York City, Philadelphia,

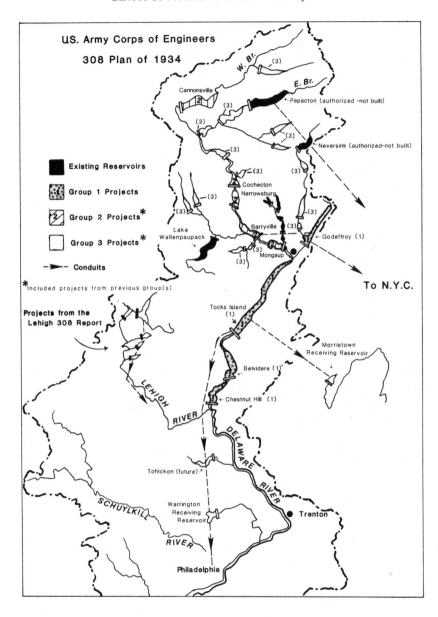

US. Army Corps of Engineers

308 Plan of 1934

W. Br. (3)

E. Br.

Cannonsville

(3)

Pepacton (authorized -not built)

(3)

(3)

Neversink (authorized-not built)

Existing Reservoirs

Group 1 Projects

Group 2 Projects*

Group 3 Projects*

Conduits

*Included projects from previous group(s)

(3)

(3)

(3)

Cochecton
Narrowsburg

(3)

(3)

(3)

Barryville (3)

Lake
Wallenpaupack

(3)

Godeffroy (1)

(3)

Mongaup

(3)

To N.Y.C.

Projects from the
Lehigh 308 Report

Tocks Island
(1)

Morristown
Receiving Reservoir

Belvidere (1)

Chestnut Hill (1)

LEHIGH

RIVER

DELAWARE

RIVER

Tohickon (future)

SCHUYLKILL

Warrington
Receiving
Reservoir

Trenton

RIVER

Philadelphia

and other cities that wanted the dam sites preserved for water-supply uses, along with various coal and private power companies that did not want competition. The Federal Power Commission rejected the application on February 14, 1936, but the company would try again in 1946, 1947, and 1949. It was never successful. The stiff opposition met by the company each time demonstrated that the Delaware River was to be a water-supply river and not a hydropower river.

One month after the first Electric Power Company of New Jersey application had been turned down, floods from two successive storms ravaged the Delaware River Basin. Along the Delaware River, flood damage was small, although it was the second worst flood on record. Floods on tributaries were more serious, prompting Francis E. Walter, the U.S. congressman from Easton, to introduce legislation to make $30 million available for implementation of the Corps' 308 plan. Although nothing happened with this proposal, the widespread flooding in the Delaware and many other areas resulted in passage of the Flood Control Act of 1936. This act expanded the federal role in flood control and assigned the responsibility to the U.S. Army Corps of Engineers. The Corps' flood-control responsibilities would make it one of the nation's largest dam-building agencies.

The flood of 1936 and one that occurred in July 1942 did little to change the perceived need for flood control on the main stem of the Delaware. Instead, the Corps' interest was directed to a long-standing project, the New Jersey Cross-State Ship Canal connecting Philadelphia with New York City. The idea of a ship canal across New Jersey had originated with William Penn many years before. The Philadelphia City Council revived the proposal in the 1890s, and in 1895 the Corps of Engineers surveyed potential canal routes for the city.

Interest in the canal increased when it was made part of the Atlantic Intracoastal Waterway, an inland waterway stretching from Boston to Florida. Primarily associated with the lobbying efforts of the Atlantic Deeper Waterways Association, the inland waterway was essentially complete by 1930 except for the New Jersey canal. The Corps of Engineers' New York District conducted studies of the canal in 1912, 1913, 1917, 1920, 1930, 1934, and 1936. In each case the benefits of the canal were found to be less than its costs. Yet the lobbying effort for the canal continued unabated.

The major water problem associated with the Cross-State Ship Canal was the need for freshwater to retard the intrusion of saltwater from the Delaware Estuary and Raritan Bay. The saltwater intrusion problem related to the depth of the canal in relation to the elevation of the two tidal rivers. One proposed solution was a dam with locks across the Delaware Estuary near Bordentown, New Jersey. Between 1936 and 1945 the New York District also studied the use of upland dams to supply adequate freshwater flows to the canal inlets. Dams at Cannonsville, New York, at Stoddardsville on the Lehigh River, on Flat Brook in New Jersey, and, in particular, at Tocks Island were investigated. Interest in the Tocks Island site led to test borings in 1942. The borings were taken down to 140 feet below the river, but no bedrock was found. Dam construction at Tocks Island was considered questionable at best and likely to be costly, if feasible.

By adjusting interest rates and benefits, the Corps finally made the Cross-State Canal look economical. Opposition to the canal was intense, particularly from the Delaware River Basin water community. Nevertheless, the canal came close to being funded by Congress in 1943. The questions about salinity and the canal's conflict with future water supplies were never resolved, however, and the canal was not built.

While the New York District was studying the canal, the Philadelphia District was reexamining its Delaware River 308 plan. The House Committee on Rivers and Harbors had directed the review in 1939 in order to determine "the advisability of constructing dams in the vicinity of Tocks Island, Belvidere and Chestnut Hill for the development of hydroelectric power, for the improvement of existing navigation facilities on the river below, and for other beneficial effects, including possible sources of public water supplies that can be made available by said dams."[6]

World War II delayed the completion of the 308 review study until mid-1946. The study turned up little that was new. Based on the New York District's boring tests, however, the Tocks Island dam of the original 308 report was rejected as "impracticable because of foundation conditions unfavorable to dam construction."[7] A dam in the Delaware Water Gap was examined as a substitute but discarded because of the extensive property, railroads, and highways that would be flooded. The Corps finally zeroed in on Wallpack Bend as the best site for a Delaware River dam. However, without an overriding flood-control or navigation need, a Corps dam on the Delaware was still premature.

By the 1930s the growing power of the federal government was alarming many states. To counter this threat, adjacent states began looking for ways to work together on common problems. Nationally, this movement was encouraged by the Council of State Governments. Early in its history, the council promoted the creation of "Commissions on Interstate Cooperation" in each state. A model commission consisted of fifteen members: five from a state's house of representatives, five from its senate, and five administrative officials appointed by the governor. The first state to organize such a commission was New Jersey in March 1935, followed by New York in April and Pennsylvania in May.

On April 3, 1936, the Subcommittee on Stream Pollution of Pennsylvania's Commission on Interstate Cooperation hosted a conference with its counterparts from New York and New Jersey. The conference resulted in the formation of the Interstate Commission on the Delaware River Basin, or Incodel, a subagency of the three Commissions on Interstate Cooperation. The State of Delaware joined in 1938. Although Incodel had no powers, the states saw it as a way to develop joint plans and programs for the Delaware River Basin. First priority on the Incodel agenda was the cleanup of stream pollution. The second was water supply.

The Incodel water-pollution control program was highly successful. The agency got water-pollution laws enacted and sewage-treatment plants built throughout the Delaware River Basin. Although inadequate by today's standards, the new waste-treatment facilities resulted in an effective first-generation cleanup effort that was in place by the late 1950s. The improvement was most noticeable in the grossly polluted Delaware Estuary, which had no municipal or industrial waste-treatment works prior to Incodel. Also completed under the aegis of Incodel was the cleanup of the massive coal-mine siltation problem in the Schuylkill River.

Incodel's involvement in water supply began with the formation of a Committee on Quantity in 1937. Early in its history, the committee disputed with Incodel because committee members wanted Incodel to develop a comprehensive water-resources plan for the Delaware River Basin. Incodel, however, wanted the committee to develop a water-allocation formula for the basin. A compromise was eventually reached through the intervention of the National Resources Committee.

Meanwhile, a new proposal to use the Delaware River came from New Jersey. In a May 1938 speech to the legislature, Governor A. Harry Moore proposed the use of the Delaware and Raritan Canal "as an integral part of our next major water supply development."[8] The canal took water from the Delaware River and carried it to the central part of the state. Ownership of the canal had reverted to the state in 1934 when it was abandoned for transportation. Shortly after his address, Moore appointed an engineering committee to study the canal proposal in detail.

Moore presented the results of the engineering study to the legislature in the following year. For $29 to $34 million, New Jersey could get 150 mgd from the canal. This was much cheaper than building dams at Chimney Rock, Long Hill, or elsewhere. Nothing was done immediately with Moore's proposal, but in 1941 the legislature funded a study of the canal by the Department of Conservation and Development and the Delaware and Raritan Canal Commission, a special commission formed in 1935. The study recommended that New Jersey develop plans for using the canal as a recreational waterway and industrial water supply. Meanwhile, the state was to keep the canal functional in order to protect its historic diversion of Delaware water.

The New Jersey proposal to expand the use of the hundred-year-old Delaware and Raritan Canal concerned Incodel greatly. The commission was irritated that the state had acted unilaterally and was not considering some flow compensation for its diversion. Meetings were held and suggestions raised. As it turned out, however, nothing came from New Jersey's proposal for another decade.

During this time Incodel's Quantity Committee was hard at work developing operating formulas for governing flow releases from future Delaware River dams. Using the principles established by the 1931 Supreme Court case, the committee came up with three control flows at Trenton: 4,000, 3,400, and 2,500 cfs. These were designed to guide reservoir operations during low-flow periods. The exact flow releases from different dam projects depended upon whether the reservoir was above or below Port Jervis.

The important feature of the formula system was that each state was free to develop its portion of the Delaware any way it wanted. As Incodel saw it: "The acceptance of these principles by the States should eliminate future controversies in regard to water supply diversions. Thus unnecessary and expensive litigation will be avoided. Moreover, providing that the participating state governments are equipped and willing to resolve their own problems in this field harmoniously, it will go a long way in precluding the necessity for federal intervention in the interstate water problems of the Delaware River Basin."[9] As this statement implies, one of the missions that had evolved out of Incodel's short lifespan was that of keeping federal government out of the Delaware River Basin.

Incodel had some justification for its antifederal bias. The 1936 flood-control act had greatly expanded the Corps of Engineers' role in dam building. At the same time, TVA-like proposals were being promoted in various parts of the country. Finally, in 1940 the U.S. Supreme Court had ruled in the New River case that federal authority extended beyond navigable waters to include all waters draining to navigable streams. To the states and Incodel it looked as if the federal government was about to usurp their own responsibilities in the field of water resources. Incodel looked upon itself as the "home rule" alternative for the rest of the nation.[10]

Incodel's attempt to establish a formula for future diversions and flow releases failed. The model diversion bill was submitted to Pennsylvania, New York, and New Jersey in 1943. The first two states quickly passed it, and New Jersey passed it in 1944. New Jersey's delay was due to changes made to the bill by the New York legislature. New York City had successfully lobbied the legislature to remove a section of the bill that would have required Supreme Court approval of all new diversion projects. New York's amendment reflected a difference of opinion among the parties to the 1931 decree. Did the Supreme Court Decree apply to all Delaware River Basin diversions or just New York City's? Incodel tried to get the Supreme Court to clarify the issue, but it was never resolved.

During the 1930s and into the 1940s, four separate water-planning activities were occurring in parallel. Three have already been discussed: the activities of Incodel, studies by the Corps of Engineers, and New Jersey's proposals for the Delaware and Raritan Canal. The fourth planning activity was a series of studies done for and by the City of Philadelphia.

During the 1930s, several new proposals had emerged for a Philadelphia mountain water supply. The first of these had been proposed in 1933 by A. Mitchell Palmer, the president of the Lehigh Valley Water Supply Company. Palmer suggested a plan to the city that would have seen reservoirs built on Bushkill, McMichaels, and Brodhead creeks in the Pocono Mountains, and on Aquashicola Creek in the Lehigh watershed. The main source of water in this proposal, however, was an intake on the Lackawaxen River several miles below the outlet of Lake Wallenpaupack. Other intakes were also proposed for Shohola Creek, Paradise Creek, and Pocono Creek. Palmer's engineers believed Philadelphia could get 353 mgd from the system.

The Lehigh Coal and Navigation Company also had ideas for Philadelphia. Based on an 1818 act of the Pennsylvania legislature, the "Old Company" claimed water rights to the Lehigh River. Beginning about 1928, the company began exploring various ways of using these rights. At first the company explored the construction of a large hydroelectric project involving the construction of various reservoirs and a coal-fired, steam-electric generating plant. The hydropower facilities were to be used during peak demand periods, with the steam plant used for base-load demands. Water for the steam plant was to come from the reservoirs, and the coal from the company's mines.

The Lehigh Coal and Navigation Company's plans included construction of reservoirs on Pohopoco Creek, a Lehigh River tributary, and on the Lehigh River upstream from the steam plant. These two reservoirs potentially yielded 370 mgd for public water supply, although the customers for this water were never defined. Between 1929 and 1931, the "Old Company" purchased land in the upper Lehigh watershed as the first step toward implementing its project.

The engineers who had developed Palmer's 1933 Pocono-Lehigh plan for Philadelphia were J. H. and W. L. Lance, two Wilkes-Barre brothers. In 1937 they proposed a new water-supply plan that combined the features of the Lehigh Valley Water Supply Company plan with the water-supply features of the Lehigh Coal and Navigation Company plan. The new plan had an estimated safe yield of 913 mgd, enough for Philadelphia and many other water users.

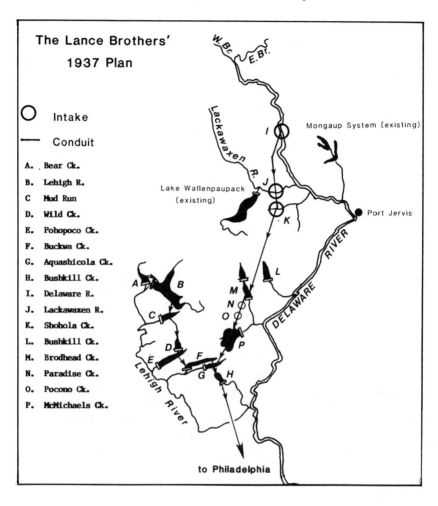

The Lance Brothers'
1937 Plan

O   Intake

—   Conduit

A. Bear Ck.
B. Lehigh R.
C   Mud Run
D. Wild Ck.
E. Pohopoco Ck.
F. Buckwa Ck.
G. Aquashicola Ck.
H. Bushkill Ck.
I. Delaware R.
J. Lackawaxen R.
K. Shohola Ck.
L. Bushkill Ck.
M. Brodhead Ck.
N. Paradise Ck.
O. Pocono Ck.
P. McMichaels Ck.

Mongaup System (existing)

Lake Wallenpaupack
(existing)

Port Jervis

to Philadelphia

In their 1937 proposal, the Lance brothers took the time to address the disadvantages of a reservoir at Tocks Island. Several of their arguments against Tocks Island Dam would be raised again and again three decades later. They included:

1. Pollution in the reservoir would mean that filtration would be required before the water could be used. Most Philadelphians assumed that Delaware River water from the upper basin could be used "as is." Undoubtedly contributing to this belief was the fact that New York City did not have to treat its water from the Catskills.
2. The Lances believed that chemicals would have to be added to the reservoir to eliminate algae.
3. The Lances feared that a reservoir at Tocks Island might destroy the recreational value of the surrounding area.
4. They pointed out that flood control at Tocks Island would be less effective than flood-control works on tributaries since these tended to flood quicker than the Delaware River.

The Lance brothers' proposal was one of several made in 1937. The Philadelphia Water Bureau also made recommendations in the same year. The bureau favored the old proposal to build reservoirs on Perkiomen and Neshaminy creeks because they were closer and thus much cheaper than a water supply in the mountains. Also favored by the bureau were a pump station on the Delaware River at Yardley, Pennsylvania, and various improvements to the Philadelphia water system.

The Lance and Philadelphia Water Bureau proposals were responses to a new study by the City of Philadelphia. A technical committee had been appointed to conduct a review of the city's water-supply alternatives. The quick study yielded nothing new, although the Corps' Tocks Island Dam idea looked particularly interesting to the engineers. Coming out of the committee's work was the recommendation that Philadelphia undertake a major engineering study of its water-supply alternatives.

Philadelphia's water survey was required to wait out World War II. After the war, a special Board of Consulting Engineers was appointed to conduct a two-part study of potential upland water sources. Part one of the survey was a preliminary reconnaissance of the various alternatives that had been kicked about for years: an intake on the Delaware at Yardley; reservoirs in the upper Lehigh watershed; reservoirs on Pocono tributaries in the upper Delaware; and a Delaware River main-stem dam project. The purpose of the first part of the survey was to decide which alternative should be examined in detail in the second part of the study. Included in the second part of the study was a detailed evaluation of the city's existing water-supply system, including an examination of the facilities needed to make it suitable for continued use. It would prove to be the largest and last study Philadelphia would make of a mountain water supply.

The Tocks Island dam site was rejected early in the study because of the geological problems noted several years earlier by the Corps of Engineers. Wallpack Bend replaced Tocks Island in the study, and it, plus two dam sites near the Delaware Water Gap and two between the water gap and Belvidere, was evaluated. Other alternatives were also evaluated, including a Lehigh-Pocono tributaries project. The first phase of the study was completed in November 1945.[11] Recommended for further study was the Delaware River dam–Yardley intake plan with Wallpack Bend as the dam site. The natural features of the dam site, plus the lack of highways, railroads, and large populations, made the Wallpack Bend site superior to all others.

In the first phase of the study, the Board of Consulting Engineers made one important change from past Delaware River dam proposals. Instead of expensive conduits delivering water to Philadelphia, water from Wallpack Bend would flow downstream to an intake at Yardley, one hundred miles away. This new way of delivering water to Philadelphia was the result of water-quality tests conducted during the study. These indicated that Wallpack Bend water would require filtration before it could be used. This eliminated the major advantage of a conduit system and was far cheaper. Withdrawing Wallpack Bend water at Yardley also meant that no water was diverted before the Delaware River reached Trenton. This was an important consideration because it avoided a potential Supreme Court fight with New Jersey.

The selection of Wallpack Bend as the potential source of Philadelphia's water was opposed by the Lehigh Coal and Navigation Company. In 1945 and 1946 the company hired consultants to develop an alternative to Philadelphia's Wallpack Bend plan. The alternative naturally favored use of the company's Lehigh River water rights. Many of the dams previously studied by the company for hydropower were resurrected and studied for water supply.

The Philadelphia Water Commission figured that the Lehigh Coal and Navigation Company was hoping to sell its "water rights" to the city. The commission therefore hired attorneys to research the legality of the company's rights. The attorneys concluded that the company's water rights were ambiguous and would probably require litigation.

In spite of the recommendations of its engineers, the Philadelphia Water Commission did not select the Yardley intake part of the project for the detailed second-phase study. Instead, the Board of Consulting Engineers was directed to study a project consisting of the dam at Wallpack Bend, a sixty-five-mile-long conduit-tunnel system, and a regulating reservoir near Warrington, Pennsylvania.

The main-stem dam detailed in the second phase of the study was to be located on the upstream bend of the Wallpack Bend "S" curve, where the Delaware River swings northeast for a short distance. The axis of the dam was therefore almost directly on a north-south line. Here the Pennsylvania side of the dam site is a steep rock cliff rising two hundred feet above the river. Across the river in New Jersey is the tip of the ridge that separates the Delaware River from the Flat Brook Valley. Unlike the Tocks Island dam site, Wallpack Bend sits on bedrock.

The dam envisioned by the Board of Consulting Engineers was to be partly concrete and partly earthen. The concrete section was the main part of the dam running from the Pennsylvania cliffs into New Jersey. In the center of the concrete section were the gates and conduits that would make flow releases or pass floods through the dam. In the Pennsylvania corner of the dam was to be the intake for Philadelphia, and in New Jersey were to be facilities for generating electricity.

The reservoir behind the Wallpack Bend dam would have been thirty miles long and about one-half mile wide in most locations. The potential storage capacity of this reservoir was estimated to be 121.5 billion gallons, with a usable storage of 91.5 billion gallons. This capacity guaranteed Philadelphia a water supply of 500 mgd. Since Philadelphia would have to filter the reservoir water anyway, recreational use of the reservoir was considered not only feasible but one of the assets of the project.[12]

In the second part of the study,[13] the Board of Consulting Engineers also analyzed the option of upgrading the city's existing water-supply system. This was found to be significantly cheaper than the Wallpack Bend dam and its expensive conduit-tunnel—$62,568,000 versus $284,588,000. The continued use of the city's existing system, however, was predicated on the cleanup of the water-pollution problems of both the Delaware and Schuylkill rivers. The cleanup effort, under the auspices of Incodel, had only recently begun, and its success was by no means assured. The board believed that the only way of guaranteeing good-quality water was to get it from the upper Delaware River Basin.

The Philadelphia Water Commission examined the final report of its Board of Consulting Engineers at length, eventually deciding not to build the Wallpack Bend dam. Instead, the city would modernize its existing filtration plants and build sewage-treatment plants. The upland water-supply dream was not abandoned, however. In 1947 the city applied to the Pennsylvania Water and Power Board for a priority in rights to the Wallpack Bend dam site. The request would have placed Philadelphia first in line for the dam site as far as other Pennsylvania communities were concerned.

By 1947 it was also clear that the Incodel water-supply diversion bills were dead. Moreover, time was running out. The recent Philadelphia interest in a Delaware River dam was seen as the first of many studies that were going to come in the postwar period. Would there soon be new proposals for private power dams, the Cross-State Ship Canal, New Jersey's canal diversion, the third phase of the New York City Delaware Project, another Philadelphia Wallpack Bend proposal, or other projects involving the Delaware River? Incodel thought so.

Incodel, which had earlier refrained from doing comprehensive water-resources planning, decided it was time to develop its own water project. James H. Allen, Incodel's chief engineer and executive secretary, had been a member of the Philadelphia Water Commision during its recent studies and personally favored a dam at Wallpack Bend. Others in the Incodel community shared his interest. To get the study started, an interstate compact was drafted that, if enacted, would have converted Incodel into a compact agency for study purposes (the Delaware River Basin Resources Development Compact). Instead, each state enacted legislation that directed Incodel to develop a multistate water project.

With state appropriations, Malcolm Pirnie Engineers of New York City and Albright and Friel of Philadelphia were hired in mid-1949. Incodel's consultants finished the first phase of their study in January 1950. In the same month, the New York City Board of Water Supply was authorized to proceed with plans to develop an additional 345 mgd from the Delaware. Incodel knew that if it failed another major water fight would occur among the states.

In August the second part of the consultants' study was completed. Incodel quickly adopted the report's recommendations. The *Report on the Utilization of the Waters of the Delaware River Basin*, or the "Pirnie Report," laid out an ambitious water-development plan for the Delaware River Basin. The plan promised new water supplies for New York City, North Jersey, the Philadelphia area, Wilmington, and areas in between.

The first phase of the Incodel plan called for the construction of a dam at Cannonsville on the West Branch of the Delaware River with a storage of 118 billion gallons. Water from the reservoir was to be released downstream to a smaller reservoir near Barryville, New York. This reservoir was to have a capacity of 10 billion gallons. During high river flows, excess water would be pumped from Barryville to a third reservoir, located in the Neversink watershed near Godeffroy, New York, for use by New York City. The cost for this part of the Incodel plan was estimated to be $517 million. It provided an assured low-flow at Trenton, New Jersey, of 4,000 cfs and water for New York City.

A dam at Wallpack Bend was an option in phase one, depending on the desires of Philadelphia and New Jersey. With a storage capacity of 121 billion gallons, the reservoir could provide up to 500 mgd of "mountain" water to Philadelphia and an additional 800 cfs to the assured low-flow at Trenton. There were two options for getting Philadelphia's water to the city. The recommended method was to release the water to the Delaware and pump it

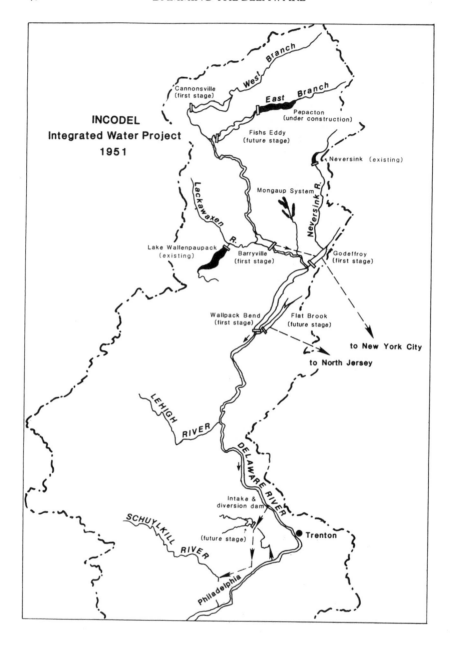

INCODEL
Integrated Water Project
1951

later to a reservoir near Newtown, Pennsylvania. A diversion dam and intake for this purpose were tentatively located upstream from Trenton. The second method, relegated to the future, was to build a pressure tunnel directly from the reservoir to a receiving reservoir near Chalfont, Pennsylvania. The price tag on the reservoir at Wallpack Bend was estimated to be $47 million.

Incodel's Wallpack Bend dam was similar to the one outlined in the Philadelphia Water Commission study. It was located at the same spot in the bend and was similar in design. Although physically similar, Incodel's Wallpack Bend dam was not exclusively allocated to water supply. Instead, a major purpose of the dam was to add water to the Delaware during times of low-flows. The difference was subtle but important. In the Delaware River Basin, water supply and low-flow augmentation were coming to mean the same thing. The two different aspects of water resources had not yet merged, but they were now much closer than before.

The second phase of the Incodel plan called for more dams and reservoirs. These reservoirs were planned for Fishes Eddy on the East Branch, Delaware River; on Flat Brook in New Jersey; and at Wallpack Bend (if not built sooner). The Fishes Eddy and Flat Brook dams increased the yield of the system by an additional 450 mgd. New Jersey's share of the system was obtained by pumping Wallpack Bend water into the Flat Brook reservoir for transfer to North Jersey.

Shortly after the Pirnie Report was released, a draft interstate water compact was completed by Incodel's compact drafting committee. The Delaware River Basin Water Commission Compact would have created a water-resources construction and operating agency with three representatives from each state. Unlike the compact agencies proposed in the 1920s, the Delaware River Basin Water Commission was to have no regulatory or planning functions.

In January 1951 Incodel submitted its plan and draft compact to the states. Each state was requested to adopt and approve the integrated project in principle, to ratify the proposed compact, and to authorize $1 million for the Delaware River Basin Water Commission's first two years of operations. Most of this money was to be spent on survey and design work. In the third year, the commission was to issue bonds for $416 million for the water-supply features of the project and $150 million for the flow-regulation features.

In early 1951 Incodel had every reason to believe that its plan would be adopted. On November 15, 1950, New York State's Water Power and Control Commission had given its blessing to New York City's application for a reservoir at Cannonsville. The city had a genuine interest in integrating this project into Incodel's basinwide system. In the lower Delaware Valley, the U.S. Steel Corporation had recently announced plans to build the world's largest integrated steel-mill complex. All along the Delaware Estuary was a belief that more industry and people would be coming soon to the region. A drought, occurring just as Incodel launched its study, had scared New Jersey water officials, and they were very interested in finding new water for North Jersey. Philadelphia, meanwhile, was becoming increasingly concerned about salinity in the Delaware Estuary, and the city had yet to begin the water-supply and waste-treatment improvements recommended by its engineers several years earlier. The Incodel plan looked like the answer to many problems.

The Incodel project was not to be, however. Adoption of the plan began well. Delaware and New Jersey quickly passed the necessary legislation in June 1951. In New York passage was assured, although the state delayed passage because of actions taken by the Pennsylvania legislature. New York did, however, enact the Incodel legislation on April 15, 1952.

As New York's hesitation indicates, opposition to Incodel's program had arisen in Pennsylvania. Some opponents believed that the plan benefited only New York City and that Pennsylvania would lose its water rights while being forced to pay for New York's new reservoir. Even before the Pirnie Report was finalized, the Pennsylvania Department of Forests and Waters had hired consultants to examine the relationship of the Incodel plan to New York City's reservoir plans. The consultants' report recommended that Pennsylvania proceed cautiously and that the state oppose any increases to New York City's allowable diversion until the Incodel reservoir system was operational.[14] Other critics found fault with the Incodel plan because it was water-supply oriented and not truly multi-purpose.

The governor of Pennsylvania, John S. Fine, stepped into the controversy by appointing a special panel to evaluate the Incodel plan in June 1951. The appointment of the Pennsylvania Water Resources Committee alarmed Incodel's supporters because the state legislature did not meet in 1952. New York City observed the events in Pennsylvania and concluded that the Incodel plan would not be approved. It then filed a petition with the U.S. Supreme Court asking to be allowed to increase its water diversion to 800 mgd. The city planned to build a reservoir at Cannonsville with or without the downstream states.

In February 1953 the Pennsylvania Water Resources Committee submitted its findings to Governor Fine. Both the Incodel integrated water project and the proposed compact were recommended for rejection. The committee believed that the primary beneficiaries of the project were New York and New Jersey since Philadelphia was still undecided about its interest in an upland water supply. The compact was rejected on the premise that Pennsylvania would be at the mercy of the other two states when it finally decided to develop water supplies in the Delaware. A safer alternative in the view of the committee was to rely on the Supreme Court to arbitrate Delaware River water disputes. The committee also felt that Pennsylvania had viable alternatives to the Incodel plan that should be explored. These included the construction of dams in the upper Lehigh and on Pocono tributaries and the building of a dam at Wallpack Bend either by Pennsylvania itself or jointly with New Jersey. None of these alternatives needed a compact or involved the state with New York City.

The report of the Pennsylvania Water Resources Committee did considerable damage to the Incodel project. Supporters of Incodel yelled foul. They noted that the committee's counsel was also the chief counsel of the Lehigh Coal and Navigation Company. Also, one of the engineering firms hired by the committee had worked for the company during the 1945–46 debates with the Philadelphia Water Commission.

By strenuous lobbying, the Pennsylvania members of Incodel and other supporters attempted to circumvent the damage created by the committee's rejection and Governor Fine's personal opposition to the project. In June 1953 these efforts resulted in passage of the Incodel bills in the Pennsylvania House. The bills, however, were never reported out of committee in the Senate. After that, the Incodel proposal faded.

The Incodel plan had stimulated one effort worth noting. The Delaware River Development Corporation was organized in late 1949 in New Jersey, undoubtedly in response to the intiation of the Incodel study. In 1951 the company received a Federal Power Commission Preliminary Permit for study of three Delaware River power dams. The largest was at Tocks Island and was to be a rock-filled dam with a twenty-two-mile-long lake and a generating capacity of 150,000 horsepower. The company planned to substitute a dam at Wallpack Bend if Tocks Island was not feasible. The other power dams were proposed for Belvidere and Chestnut Hill, power-dam sites that had been talked about for years. The Delaware River Development Corporation renewed its permit annually until 1954, when the Federal Power Commission refused to renew it any longer. The corporation therefore dissolved on January 27, 1955. It was resurrected later, however, when Mother Nature interceded in the water affairs of the Delaware River Basin.

As for Incodel, its integrated water project and interstate compact would die a slow death in Pennsylvania. Incodel itself would survive until 1962, although its influence was greatly diminished. The dream of a dam across the Delaware River governed by an interstate compact would not die, however. New people, assisted by another interstate water fight and a devastating flood, would step forward to promote both. Tocks Island was also not dead as a dam site. Although it had been rejected twice by the Corps of Engineers, once by Philadelphia, and not even considered by Incodel, it was in the shadows, waiting to rise again.

5

The Stage Is Reset

While the Incodel plan was being defeated in Pennsylvania, action shifted to the U.S. Supreme Court. On April 1, 1952, New York City petitioned the Court to reopen the Delaware River Case. The city sought an increase in its diversion from 440 mgd to 800 mgd. Plans for a reservoir on the West Branch of the Delaware River at Cannonsville, New York, had already been drawn up by the city's Bureau of Water Supply. Internally, attempts to have the city develop a Hudson River water supply instead of another Delaware River Basin dam had been defeated. New York State joined the city's petition in late April, and shortly afterward both New Jersey and Pennsylvania responded. They planned to contest New York City's plans. On June 9 the Court appointed attorney Kurt F. Pantzer of Indianapolis as its special master.

The second Delaware River Case was much different from the first one. In 1929 New Jersey had raised the biggest objections to New York City's plans in the Delaware River Basin. This time, however, New Jersey not only accepted New York's diversion but wanted its own water diversion. The principal opposition to both New York's and New Jersey's plans would come from the Commonwealth of Pennsylvania. The first case had also laid down the principles governing water diversions from the Delaware River Basin and had established that New York City had water rights in the basin. The second case would be argued with these principles and water rights as givens.

Pantzer initiated his investigations by sifting through the voluminous testimony of the 1931 case. He then met with the attorneys of each participant and conducted a prehearing inspection tour of the Delaware River Basin. With this background investigation completed, pretrial hearings were held. The objective of the special master's preliminary efforts was to see if the three states and New York City could negotiate a settlement and thus avoid a Court-dictated decree. In early 1953 this effort bore fruit when New Jersey and New York jointly developed the so-called Montague Formula as a possible solution.

Under the Montague Formula, New York City would be allowed to increase its water diversion to 800 mgd and New Jersey would get a diversion of 250 mgd via the Delaware and Raritan Canal. New Jersey's water would be sent to the Raritan River Basin in the central part of the state. Under the proposed formula, New York City planned to release enough water from its reservoirs so that a specified minimum flow was maintained in the Delaware River below Port Jervis (i.e., at Montague, New Jersey).

Pennsylvania's response to the Montague Formula established the conditions for further negotiations. Specifically, Pennsylvania wanted:

1. New Jersey's repeal of the 1783 bi-state treaty that banned dams on the Delaware River.
2. Passage of legislation by New Jersey that would commit the state to condemning land in New Jersey if Pennsylvania decided to build a dam at Wallpack Bend and/or diversion dams at Yardley or elsewhere.
3. New Jersey's cooperation in working out a flow-release formula for a reservoir at Wallpack Bend or reservoirs on tributaries in either state. (These negotiations were never completed.)
4. A provision in the new decree making it clear that New York City's diversion was not to be considered a "prior appropriation." Regardless of the expenditures made in building Cannonsville Reservoir, Pennsylvania wanted the right to seek a reduction in the city's allowable diversion if it felt its own water needs warranted it.
5. The Supreme Court's appointment of a permanent "river master" to administer the amended decree. Pennsylvania had made the same request in the first Supreme Court case but had been denied.

In October 1953 New York City formally changed its original petition to include the reservoir release strategy contained in the Montague Formula. With certain qualifiers, the Montague Formula called for the maintenance of a flow at Montague, New Jersey (actually Milford, Pennsylvania), of 1,525 cfs before Cannonsville Reservoir was completed and 1,750 cfs afterwards. The qualifiers gave conditions when the flow requirements could be altered. Pennsylvania objected to the qualifiers, contending that the flow objectives should be absolute.

Formal hearings were held from October 15, 1953, to April 1954. In the last days of the hearing process, the State of Delaware petitioned to intervene in the case, and this was granted by the Court. Behind the scenes, negotiations were still occurring. When all the parties to the case were in agreement, the special master prepared a report and recommended a decree for the Court. The U.S. Supreme Court adopted this decree on June 7, 1954, without rendering an opinion.

The provisions of the Amended Decree are crucial to an understanding of the Delaware River Basin. They allowed New York City to increase its out-of-basin water diversion from 440 mgd to 490 mgd when the construction of Pepacton Reservoir was completed. This reservoir was the large East Branch reservoir that had been authorized by the 1931 decree and was now almost finished. When New York City completed the construction of its reservoir at Cannonsville, its diversion rights were to be increased to 800 mgd. There were no conditions placed on New York City's use of this water, meaning that the city could use it wisely or waste it if it wanted. Like the 1931 decree, the Amended Decree contained wording to the effect that the city's diversion was not necessarily granted forever.

Flow-release requirements were of the utmost importance to the downstream states. When Pepacton Reservoir became operational (on September 1, 1955), the Montague Formula required releases from New York City reservoirs that were sufficient to maintain a flow at Montague of at least 1,525 cfs. This minimum flow increased to 1,750 cfs when Cannonsville Reservoir came on line (on March 31, 1967). The city also had other flow-release requirements. On an annual basis, "excess" storage releases were to be made—a quantity of water equaling 83 percent of the difference between actual water consumption and the safe yield of the city's water-supply system. This provision ensured that the city would not keep unused water locked up in its reservoirs. Instead, it would be released to the downstream states.

Under the Amended Decree, New Jersey received permission to divert 100 mgd (as a monthly average) from the Delaware River Basin. This authorization, less than it had requested, formalized the state's historic Delaware and Raritan Canal diversion and placed the diversion under the jurisdiction of the Court. Because of the historic nature of the canal diversion, New Jersey was not required to meet any flow-compensation requirements. It was understood, however, that diversions higher than 100 mgd would be contingent on the state's providing compensating flow releases from one or more upstream reservoirs.

During the negotiations, New Jersey had satisfied one of Pennsylvania's conditions. On December 30, 1953, legislation was enacted that granted Pennsylvania the authority to build a dam at Wallpack Bend or diversion dams such as the kind that would be needed for an intake at Yardley. New Jersey also agreed to exercise the power of eminent domain on Pennsylvania's behalf when the latter moved to purchase land in New Jersey. New Jersey, on the other hand, was allowed to buy into the Wallpack Bend dam project for as much as 30 percent. Finally, the New Jersey legislation revoked the antidam provision of the Pennsylvania and New Jersey Compact of 1783. Since the New Jersey legislation took effect when Pennsylvania passed reciprocal legislation, passage of the reciprocal legislation was made part of the Amended Decree.

The New Jersey legislation was, of course, part of Pennsylvania's spoils in the Supreme Court case. Pennsylvania also got its permanent river master, the chief hydraulic engineer of the U.S. Geological Survey. His job was to administer the Amended Decree, particularly the technical aspects of the Montague Formula. With a river master in charge of the flow releases on a daily basis, Pennsylvania felt its needs would be protected.

In the aftermath of the June 1954 Supreme Court decision, *Engineering News-Record*, a leading engineering magazine, asked "why a master for the Supreme Court succeeded in doing what the Interstate Commission on the Delaware River Basin failed to accomplish."[1] The magazine's question generated two responses. Kurt F. Pantzer suggested that Pennsylvania and Delaware were not yet ready to use the water that the Incodel plan would have created. The other response was more pragmatic. According to Samuel S. Baxter of the Philadelphia Water Commission, "If the matter at issue is the solution of the problem of the use of the Delaware River, then I believe that the Supreme Court decision has not settled the matter."[2] He illustrated his argument by pointing out that one of Pennsylvania's expert witnesses had fixed 1978 as the year in which the needs of the lower basin would no longer be met. Baxter therefore assumed that the matter would be back to the Supreme Court in fourteen years.

Engineering News-Record discussed the Delaware River Basin situation in a subsequent editorial and suggested that the next step should be taken by Pennsylvania:

Only Pennsylvania balked at Incodel's proposed compact before. It would be an admirable thing indeed if Pennsylvania were now to assume the role of responsibility; to revive official interest in negotiation toward final, equitable settlement of the Delaware issue; and resume very intensive public education in the water supply problem it will be up against in 14 years. Confronting the problem now, solving it, and insuring the solution with that higher sanction—the interstate compact—will mean avoiding future expedient court action. Such a step, too, will allow much more time for thorough, unhurried engineering planning and design of future projects on the Delaware.[3]

The role envisioned by *Engineering News-Record* for Pennsylvania was assumed by Philadelphia. Indeed, the magazine's editors may have been echoing the thoughts of Philadelphia's mayor and his advisors. Mayor Joseph S. Clark was very interested in having someone take a fresh look at the Delaware, and he personally favored the establishment of a TVA-like agency in the basin.[4] In an August 9, 1954, letter to Clark, Samuel Baxter outlined various alternative steps that the city could take. One alternative was for Pennsylvania to build one or more dams across the Delaware, probably at Wallpack Bend. This alternative, however, involved huge expenditures by the state. Opponents of the Incodel plan in Pennsylvania had made much of the money that had been saved by the plan's defeat. If Philadelphia built the dam, the city's leaders would undoubtedly face the same opposition. Instead, Baxter recommended that Philadelphia "call for a complete study of the problem in light of the Court decision. This should be made by regular state agencies, responsible to the Governor and the Legislature, and looking toward the time when Pennsylvania must take further action. Such study would be useful no matter whether the case is handled in the future by an interstate compact or by reopening the Court decision."[5] In the meantime, he recommended that Philadelphia pressure the Pennsylvania legislature to pass the reciprocal legislation repealing the 1783 antidam treaty provisions.

Philadelphia's role as a catalyst was made easier by the fact that Pennsylvania and New Jersey were still very interested in the Delaware. If nothing else, the recent Supreme Court decision had brought the two states closer together on a possible dam project at Wallpack Bend. Both states were experiencing postwar growth that would require water, and New Jersey had also never solved the water-supply problems of North Jersey. The Delaware still looked like the ultimate solution for each state's water needs. The recent Incodel plan had shown that ample water supplies could be developed there.

Mayor Clark's job, then, was to get the Delaware River Basin states, Philadelphia, and New York City together to discuss the comprehensive study being kicked around city hall. This job was made possible because, for the first time in years, the governors of the four basin states, plus the mayors of New York City and Philadelphia, belonged to the same political party. On a more local level, Clark wanted to see Pennsylvania move toward the building of a state dam at Wallpack Bend.

In February 1955 Governor Robert Meyner of New Jersey traveled to Philadelphia's city hall to meet with Mayor Clark and his staff. In the following month, Clark and his staff traveled to Harrisburg to meet with Pennsylvania's Governor Goerge M. Leader and his staff. Two days later, Governor Meyner of New Jersey and his staff also traveled to Harrisburg for a meeting with Pennsylvania's governor. When all the meetings were over, Mayor Clark's proposed study of the Delaware was much closer to being realized.

While the various meetings were occurring, Vernon D. Northrop, Philadelphia's director of finance, and Paul N. Ylvisaker of the mayor's office were preparing staff studies concerning the pending Pennsylvania legislation to repeal the 1783 treaty, the proposed comprehensive planning activity, and the reasons the Incodel plan had failed. These were to be used as the basis for discussion. Meanwhile, both were in contact with Resources for the Future, a Washington, D.C.–based research organization. Resources for the Future had been the recipient of a $3.5 million Ford Foundation grant and was very interested in river-basin planning. The mayor's staff looked to the organization for the funding for their basin planning, and Resources for the Future was interested in using the Delaware as a potential prototype study.

After the initial meetings between New Jersey and Pennsylvania, it was time to meet with New York. A contingent from each state traveled to New York City in late April to meet with New York State water officials. The purpose of this meeting was to lay the groundwork for a major water summit the following month. This was held on May 17, 1955, at the Nassau Tavern in Princeton, New Jersey. Each of the Delaware River Basin states, including Delaware, sent the heads of its water and commerce departments to the meeting. Philadelphia and New York City sent similar high-level staff, and representatives of Incodel and Resources for the Future also attended.

The discussion at Princeton was straightforward and wide-ranging. Most parties agreed that the Incodel plan was now dead and that a reexamination of past water studies would not hurt. The new study was not to be an engineering study but a look at possible subjects for future engineering or research. If possible, the first effort was to yield an outline of a comprehensive plan. The conferees agreed to establish a temporary committee of nonpublic officials to conduct the study. Each state and city would appoint one citizen to serve on the committee. Funding for a small study staff and other expenses was anticipated from Resources for the Future.

The idea of using a committee of nonpublic officials had originated with Mayor Clark. The mayor wanted someone to take a fresh look at the Delaware River water problems, and he felt that the existing water bureaucracies could not deliver an unbiased and innovative evaluation. The use of Incodel, in Clark's mind, was fraught with particular danger since it was still fighting to have its own plan adopted. The committee envisioned by Clark would consist of close confidants of each governor and mayor and thus would reach the highest level of policymaking.

While this was going on, other planning activities were being pursued in both Pennsylvania and New Jersey. In New Jersey, a seven-member Legislative Committee on Water Supply had been created on January 1, 1955. The commission's job was to conduct a study of the state's water-supply alternatives, including a "Delaware River Valley Water Supply Project." Consultants submitted a preliminary report to the commission in July recommending that the state build a reservoir at Chimney Rock near Bound Brook. The reservoir would be adequate until 1995, when additional water was to be acquired from the Delaware River. A state dam-building agency was also recommended, but in November 1955 the voters of New Jersey rejected both it and a bond issue containing funding for the Chimney Rock project.

Pennsylvania, meanwhile, was examining a state-built dam at Wallpack Bend. On May 25, 1955, the Department of Forests and Waters hired Albright and Friel, Inc. to review the design of Incodel's Wallpack Bend dam proposal and update the costs to 1955 prices. This firm had helped develop the Incodel plan, and one of its principals, Francis S. Friel, had been a member of Philadelphia's 1946 Board of Consulting Engineers. Pennsylvania also wanted the dam's potential recreation, flood-control, and power benefits studied. The study would be the most complete ever made of a dam at Wallpack Bend.

By August 1955 it appeared that a clear track was being readied for the construction of a dam on the main stem of the Delaware. In June 1955 Governor Leader of Pennsylvania had signed the reciprocal legislation passed earlier by New Jersey. The passage of the legislation in Pennsylvania marked the end of the antidam provision in the 1783 treaty. Albright and Friel's report on Wallpack Bend was nearing completion. New Jersey also had consultants working on its water needs, and it was quickly becoming evident that New Jersey might be Pennsylvania's partner at Wallpack Bend. Efforts to establish an interstate/city special-committee study of the Delaware River Basin were also nearing completion, and private funding for the study looked promising. Certainly such a study would point out the need for new water-storage reservoirs. One event, an act of nature, would alter all these plans.

The summer of 1955 had started dry. In fact, very little rain fell in the period from June 28 to August 11. July was a scorcher, with temperatures in the nineties on most days. The heat and lack of rain began worrying water-supply officials. Reservoir levels were declining drastically throughout the East. The region was in the beginning stage of a severe drought.

The drought ended quickly. In the afternoon of August 12, Hurricane

Connie hit the coast of North Carolina. During the next forty-eight hours, Connie would dump tremendous amounts of water in the drought-stricken eastern states. By the time the storm reached the Great Lakes, more than forty people had died. In the Delaware River Basin, rain from Hurricane Connie varied from 4 inches to 12.5 inches. Still, because of the dry soils and low streams, flooding was localized and minor. Groundwater and reservoir levels recovered, marking the end of drought conditions. Days of oppressive humidity followed Connie. The rains had thoroughly saturated the soils throughout the East.

While Connie was disappearing in Canada, another hurricane came up from the south Atlantic and hit North Carolina. Coming as it did so soon after Connie, Hurricane Diane was watched with some concern. By August 18, however, it appeared that Diane was quickly losing her punch. The Weather Bureau predicted that the storm would track north to Canada, leaving portions of western Pennsylvania, West Virginia, and western Maryland wet. The forecasters were wrong. On August 18 Diane suddenly veered east and followed a direct course to Trenton and the New England coast. Picking up moisture from the Atlantic Ocean, Hurricane Diane revived, and torrential rains fell again on the Delaware River Basin and other eastern river basins.

In many areas, the amount of rain received from Diane was actually less than that received from Hurricane Connie. Diane's rain, however, fell on soils still saturated by the heavy rains of six days earlier. Runoff was extremely heavy since there was simply no place for all the water to go. Aggravating the situation was the path followed by Hurricane Diane. Heading east, Diane's band of heavy rain crossed the major rivers perpendicularly. This resulted in maximum flooding on the larger rivers. The exception to this was the Delaware River Basin. There the hurricane followed a path that was more closely parallel to the Delaware River. Because of this, maximum flooding occurred on both the Delaware River *and* its tributaries.

The torrential rains on top of the water-saturated soils caused disastrous flooding throughout the Delaware River Basin. In the Pocono Mountains and the upper Lehigh, Diane dumped eleven inches of rain. Flooding there was extreme. On the few gaged streams, stream-flow measurements were two to five times greater than any ever recorded.[6] On Brodhead Creek, floodwaters destroyed a summer camp, killing thirty-seven women and children, and flooding in the watershed accounted for one hundred deaths (ninety-nine by some counts), or about half of the total deaths attributed to Diane's passage through the eastern United States.[7]

On the main stem of the Delaware River, no lives were lost. Flooding, however, was severe. At each of the stream-flow gages on the Delaware, new records were set on August 19 and 20. The highest flow was observed at Riegelsville, below Easton, when 340,000 cfs passed downriver on August 19. On August 20 the Delaware River crested at Trenton with a flow of 329,000 cfs. This flow was 12 percent greater than any flow ever seen at the Trenton gage.

The flood of 1955 caused about $500 million in property losses in thirteen states. Approximately 20 percent of these losses occurred in the Delaware River Basin. One week after the flooding, Representative Francis Walter of Easton called on Governor Leader of Pennsylvania and Governor Meyner of New Jersey to urge their legislatures to approve flood-control dams on the Delaware River. In his telegrams, Walter urged the construction of the long-proposed dams at Chestnut Hill, Belvidere, and Wallpack Bend for flood control.

Walter was not the only person now thinking about the need for flood control in the Delaware River Basin. Before the flood debris was cleared, Pennsylvania officials had Albright and Friel's Wallpack Bend dam report on their desks. The $70 million dam outlined in the report provided flood-control, recreation, power, and water-supply benefits. Behind the dam was to be a twenty-five-mile-long lake containing 121 billion gallons of water, four or more bathing beaches, and power facilities worth between $364,000 and $379,000 per year. Quick calculations by the engineers revealed that flood control built into the reservoir pool could reduce a flood similar to the one of August by five feet. In conjunction with the New York City reservoirs, the dam also promised a minimum flow in the Delaware at Trenton of 3,400 cfs during a drought such as that of 1929–31, the worst on record. A dam at Wallpack Bend promised much at any time. Coming right after the most devastating flood in Delaware River history, it looked particularly good.

The flood of 1955 would prove to be a major turning point for the Delaware River Basin, but not for the reasons often cited. The flood did not stimulate interest or planning for a Delaware River dam. The long-standing efforts of Philadelphia, Pennsylvania, and New Jersey were being pursued vigorously even as the first raindrops fell. The twin storms Connie and Diane did demonstrate that Mother Nature could, on rare occasions, deliver a combination of natural events that could cause massive flooding. This new factor was significant because it opened the door to massive federal involvement in the Delaware River Basin.

6

The Pieces Are Put Together

After the devastating 1955 flood, the desire to dam the Delaware was greater than ever. Flood control was clearly one of the responsibilities of the federal government, and its participation, and even leadership, in the basin was now welcome. Involvement of the federal government meant the U.S. Army Corps of Engineers, one of the nation's biggest dam-building agencies.

The Corps of Engineers had never been totally absent from the water affairs of the Delaware River Basin. Most of its nonmilitary activities, however, had been largely limited to navigation projects such as channel dredging, the Chesapeake and Delaware Canal, and harbor improvements. Heavy flooding in 1942 on the Lehigh and Lackawaxen rivers had resulted in the first authorized flood-control dams in the Delaware River Basin. Nobody disputed the Corps' involvement in these projects. The Lehigh Project, authorized in 1946, consisted of local flood-control works in the Allentown-Bethlehem area and the construction of Bear Creek Reservoir, sixty miles up the Lehigh River. The Lackawaxen River Project, authorized in 1948, consisted of two reservoirs above Honesdale and Hawley, Pennsylvania: Prompton Reservoir and Dyberry Reservoir (General E. Jadwin Dam).

The lack of federal involvement on the main stem of the Delaware River was due to the lack of serious flooding prior to 1955. The 308 report of 1933 and subsequent reviews could never find any justification for federal dams above Trenton. The last of these reviews had been authorized in 1950 by the U.S. Senate Committee on Public Works, largely at the request of

Incodel. Incodel wanted federal dollars to help its integrated water project, and the Corps was directed to review the 308 report "with the view of determining the extent and magnitude of the benefits, which are now in the National interest, of water projects recommended in the Incodel plan."[1] The Corps' review took five years, reaching the chief of engineers' desk on July 7, 1955. In one of the ironies of history, the review concluded that nothing had changed to now justify federal involvement in flood control on the Delaware River. Six weeks later, with flood damages over $100 million, the chief of engineers returned the 308 review document to the Philadelphia District for reconsideration.

The return of the 308 review report to the Philadelphia District marked the beginning of the largest study of the Delaware River Basin ever conducted. This was not apparent at first. Philadelphia's Vernon Northrop, for example, believed that the restudy would be limited, leading to appropriations for the Lehigh and Lackawaxen flood-control projects, which were authorized but not yet funded.[2] Northrop, however, was wrong. Sentiment was for a new, thoroughly comprehensive study of the basin's water resources.

Although the 1950 resolution of the Senate Committee on Public Works provided the original authorization for the Corps' study, further resolutions were forthcoming in the wake of the August flooding. On September 14, 1955, the Senate Public Works Committee adopted a resolution calling for the review of all past Delaware River Basin reports "in view of the heavy damages and loss of life" caused by the August flood.[3]

During the period September 1955 to January 1956, various public hearings were held in the Delaware River Basin in order to obtain information on the extent of flood damages and to solicit opinions concerning the need for water projects in the Delaware River Basin. At all the hearings, local and state interests were unanimous in their desire for the federal government to take a lead role in the water affairs of the Delaware River Basin. Even Incodel called on the Corps of Engineers to develop a comprehensive multipurpose program covering all aspects of Delaware River Basin water resources.[4]

On February 20 the Senate Committee on Public Works adopted another resolution concerning the pending survey. The resolution directed the Corps to examine "the feasibility of the construction and operation of a reservoir on the Main Stem of the Delaware River above the Delaware Water Gap near Wallpack Bend or Tocks Island, on a cooperative basis by the United States and the Commonwealth of Pennsylvania and the State of New Jersey."[5] The special Tocks Island study had originated in the Pennsylvania legislature. For the politically sensitive Corps, a dam on the Delaware was practically a study "given."

In 1956 the Corps' Delaware River Basin survey got under way in earnest, and the Philadelphia District assembled a special staff for the planning activity. The Valley Report Group was heavily engineer-oriented but also included a wildlife biologist, an agricultural economist, a geologist, and participants from several other disciplines. Much of the work on the basin plan, however, was farmed out to other federal agencies. Some of the more important studies included the Office of Business Economics' economic-base survey, the U.S. Geological Survey's groundwater and geological studies, the U.S. Department of Agriculture's rural water use and agriculture studies, the Federal Power Commission's studies of power markets and hydroelectric-power benefits, the U.S. Public Health Service's water-quality and water-use studies, the National Park Service's recreation studies, and the Soil Conservation Service's and U.S. Geological Survey's studies pertaining to sedimentation. The Corps of Engineers, along with its overall plan responsibilities, had technical responsibility for plan components dealing with hydrology, flood damages, flood control, navigation, water needs, costs and benefits, and hydropower feasibility.

As studies go, the Delaware River Basin survey was large and uniquely comprehensive. It was the Corps of Engineers' first truly comprehensive river-basin planning endeavor. There was little doubt from the beginning that the survey would recommend one or more dams for the Delaware River. The Corps, having a few dam-building aspirations of its own, was well aware of the political interest in such a project. Pennsylvania, in fact, was still moving ahead with its own plan for a state-built dam at Wallpack Bend. While the Corps' study was getting under way, the Commonwealth of Pennsylvania was holding hearings on its dam plans and purchasing land in the Minisink Valley. A $500,000 appropriation was used to purchase sixteen properties.

In February 1957 the Corps released the results of its special Tocks Island study.[6] The study indicated that a dam could be built at Tocks Island if an earth-fill dam structure was used. A reservoir at Tocks Island was found to have twice the effective water storage as one formed by a dam at the Wallpack Bend site. This was not only because a dam at Tocks Island was eight miles further downstream, but also because the reservoir would flood the Flat Brook Valley in New Jersey. The additional storage at Tocks Island cost only about 50 to 60 percent more; thus, it was seen as a better value. Wallpack Bend was dead as a potential dam site.

The Corps' major problem in releasing the special Tocks Island study was containing the enthusiasm of the basin's water professionals. The Corps' report addressed only Tocks Island's practicability in comparison to a similar dam at Wallpack Bend. It had yet to examine the dam's economic feasibility and its relationship to other potential dam projects in the basin. In spite of this, some clamored for the immediate construction of a dam at Tocks Island.

Francis A. Pitkin, Incodel's chairman, called for the initiation of construction in 1958.[7]

Meanwhile, the defunct Delaware River Development Corporation had risen again. Reorganized on September 2, 1955, the company had filed for another Federal Power Commission Preliminary Permit to study hydropower dams at Tocks Island, Belvidere, and Chestnut Hill. Opposition to the company's plans was intense. By mid-1957 the company was beaten, and it formally withdrew its application. The company, formed initially at the beginning of the Incodel study and then again after the 1955 flood, was probably hoping to be the private power developer in any public dam project built on the Delaware. Publicly operated power facilities were very controversial since they put the federal government in direct competition with existing private utility companies.

Mayor Clark's idea of a committee of politically influential citizens to study the Delaware had not died with the start of the Corps' large study. The Delaware River Basin Survey Commission was organized in February 1956, and each state and city allocated funds to support a small professional staff, which at its peak included Walter M. Phillips, the executive secretary; W. Brinton Whitall, a planner; and Blair T. Bowers, an engineer. The staff offices were located in Philadelphia, several floors above the offices of Incodel.

One of the first jobs of the Delaware River Basin Survey Commission involved the decision to change its name. The name was changed to the Delaware River Basin Advisory Committee (DRBAC) to avoid confusion with the Interstate Commission on the Delaware River Basin or the Delaware Basin Survey Coordinating Committee, an interagency committee established by the Corps to help coordinate its study. In its short history, the work of the Delaware River Basin Advisory Committee would be equal in importance to the Corps' large study.

With the Corps' study now recognized as a major undertaking, the DRBAC abandoned the idea of conducting its own survey of the Delaware River Basin. There was still a need for the committee, however. Somebody would have to ensure that the Corps' study was complete and modern in its approach to the Delaware. The watchdog role was part of the committee's effort. At the same time, the committee hoped to conduct research into areas that would complement and augment the Corps' study. Finally, the DRBAC hoped to establish a basinwide citizen organization that could be used to counter any opposition to Delaware River Basin dams. The DRBAC wanted to ensure that any dam project recommended by the Corps was actually built.

The immediate problem facing the DRBAC, once its new roles were defined, was its failure to get sufficient funding for its planned activities. The committee did manage to get a grant of $7,500 from Resources for the Future to fund the design of an economic-base survey for the basin. The design, prepared by the Institute for Urban Studies of the University of Pennsylvania, resulted in the expansion of the work of the Office of Business Economics for the Corps.

For major funding, the DRBAC went to the Ford Foundation. Since the Ford Foundation did not make grants to government entities, the DRBAC created a nonprofit corporation called Delaware River Basin Research, Inc. On March 25, 1957, the Ford Foundation awarded this corporation $131,000 for a study of potential administrative organizations for interstate river basins. The emphasis of the study, of course, was on the Delaware River Basin and its problems. Delaware River Basin Research, Inc. awarded the contract for the study to the Maxwell Graduate School of Public Administration of Syracuse University.

The second major effort of the Delaware River Basin Advisory Committee was the establishment of a public-information program concerning Delaware River Basin water resources. In 1957 a steering committee was established to develop such a program. One of the committee's recommendations was the establishment of a "nonprofit, impartial" educational organization.[8] The vehicle for the establishment of the organization was again Delaware River Basin Research, Inc. In May 1959 the corporation was renamed the Water Research Foundation for the Delaware River Basin. A campaign to solicit funds from business and industry was conducted and more than $100,000 was raised. In 1959 the foundation produced a public-information pamphlet entitled *Water for Your Ever-Expanding Needs.*

The recommended basinwide citizen's organization was accomplished on May 22, 1959, when representatives of two hundred or so groups formally organized the Water Resources Association of the Delaware River Basin (WRA/DRB). The organization had actually been incorporated three months earlier. An executive director, Frank Dressler, was hired. In the first year of its existence, the WRA/DRB, with the help of the Water Research Foundation, developed a mobile exhibit that traveled to various locations, published six widely disseminated newsletters, and produced two booklets, *Too Much or Too Little* and *Water for Recreation—Today and Tomorrow,* two television commercials, and filmstrips. In addition, members of the WRA/DRB speakers bureau gave talks at various locations in the basin and testified at Senate hearings. With the help of the WRA/DRB and others, area newspapers began publishing articles on the Delaware River Basin. The publicity had a steamroller effect that generated excitement, and presumably public support, for water-development projects in the Delaware River Basin.

The focus of the publicity in 1959 was on two events. In January 1959 the

Corps of Engineers had publicly announced its preliminary reservoir plan for the Delaware River Basin. Five dams were recommended for construction before 1980. The objectives were to reduce flood damage in the Delaware River Basin, to support eight million annual "visitor-days" of recreation, and to assure a minimum flow in the Delaware River of 3,400 cfs at Trenton and 600 cfs in the Schuylkill River near Philadelphia. Beyond 1980, sixteen additional dams and reservoirs were believed to be necessary. The keystone of the whole program was a dam at Tocks Island.

The second event of 1959 was the completion of the Syracuse study. The report, entitled *The Problem of Water Resources Administration with Special Reference to the Delaware Basin*, was delivered to the Water Research Foundation on September 1, 1959, and released to the public shortly thereafter. It was later published in book form under the title *River Basin Administration and the Delaware.*

The bottom line of the Syracuse research was the recommendation that a water-resources agency be established by compact for the Delaware River Basin. This finding was hardly new. What was new was the recommendation that the federal government be a partner in the compact arrangement with the four Delaware River Basin states. Such an arrangement had never been tried before. The agency to be created by the recommended compact, the Delaware River Commission, was to have broad powers within a framework of existing federal, state, and local agencies. The commission was not to supplant them but to provide overall planning, coordination, and supervision. Most water-resource functions were assigned to the agency, including water-pollution control, the construction and operation of dams and other structures, the regulation of water withdrawals and diversions, research and data collection, and the review and control of water projects developed by both government and private interests. Revenues for the agency were envisioned as coming from the five members, as well as from the sale of water, the sale of hydroelectric power, and charges for recreation. In addition, the commission could issue bonds to finance anything it wanted. This, plus the commission's regulatory powers, made it potentially very powerful.

On September 30, 1959, the governors of New Jersey and Pennsylvania, representatives of the other two Delaware River Basin governors, and the mayors of Philadelphia and New York City met to review the Syracuse findings. The group directed the DRBAC to "prepare a proposed draft of legislation for the creation of a basin agency by interstate-federal compact."[9] In calling for the development of the draft compact, the governors and mayors rejected a Syracuse recommendation for an interim agency. The so-called Delaware River Agency for Water was to be federally created, with state representation. Such a step was seen as time-consuming, with no guarantee that the interim agency would evolve into a permanent interstate-federal compact agency.

The actual drafting of the interstate-federal compact was largely the responsibility of William Miller, a noted Princeton attorney hired by the DRBAC to be its legal consultant. Miller eventually prepared four drafts of the proposed compact. At each step in the drafting process, assistance and advice were obtained from the DRBAC members and a special committee of state and city attorneys. The process lasted through 1960.

Among the major issues resolved during the negotiations was the question of membership on the compact commission for New York City and Philadelphia. Both voting and nonvoting memberships were debated. Another question was whether each commission member should have one or three representatives. Eventually it was decided that each party to the compact would provide one commissioner. In the case of the states, the representatives would be the governor; for the federal government, the representative would be an appointee of the president. The two cities would be advisors to their respective states but not members.

Another important issue was whether public power interests (municipally owned interests, rural cooperatives, etc.) should be given preference for any power obtained from the publicly owned dams in the basin. New York State favored a preference clause, but it was opposed by factions in both Pennsylvania and New Jersey. One reason for this opposition was the belief that power sales would be the major revenue source of the new commission. Power sold to private power companies could command a higher price than power sold to public entities. The issue was resolved by not resolving it. Instead, the question was put off to the day when dam contracts were to be negotiated.

The thorniest issue during the negotiations was what to do about the U.S. Supreme Court Decree. The relationship of the decree to the compact was critical since the compact commission would have power over water diversions and reservoir operations. In addition, it was conceivable that the commission could condemn and assume ownership of New York's reservoirs. New York City wanted protection. Also confusing the issue were New Jersey's diversion rights and Pennsylvania's right to reopen the Supreme Court case when it felt threatened.

After many months of debate, the compact drafters agreed to freeze the diversions and flow releases authorized by the 1954 Supreme Court Decree. The freezing of the decree provisions meant that New York City's 800 mgd diversion was now permanent and not subject to future arbitration. The compact commission, however, would have power over new diversions and flow releases. After this issue was successfully resolved, the negotiations were completed and the compact language finalized to everybody's satisfaction.

This set the stage for the adoption of the compact by each Delaware River Basin state.

On February 1, 1961, the sponsors of the Delaware River Basin Advisory Committee, the four state governors and the two city mayors, met in Philadelphia. Attending were Governor Nelson A. Rockefeller of New York, Governor Elbert W. Carvel of Delaware, Governor David Lawrence of Pennsylvania, Mayor Robert F. Wagner, Jr., of New York City, Mayor Richardson Dilworth of Philadelphia, and a representative of Governor Robert B. Meyner of New Jersey. The six men endorsed the final version of the compact and promised to submit the compact legislation to their respective legislatures.

Before nightfall the compact legislation had already been introduced in the New York legislature. Support for the compact in New York State was widespread. New York City had the most to gain since the compact guaranteed its water diversions, something the Supreme Court Decree did not. The compact bill passed the New York State Senate by a vote of 57 to 0 on March 14, 1961, and the State Assembly on the following day by a vote of 145 to 2. On March 17 Governor Rockefeller signed the compact into law.

In New Jersey, action on the compact legislation was delayed for a short time. Sussex County officials wanted time to study the effects of the compact on their county, including the impact of building Tocks Island Dam. Several business groups also took time to study the proposal. These concerns were soon overcome. The New Jersey Senate passed the legislation on April 24 by a vote of 16 to 0, and the House followed suit on May 1 with a vote of 51 to 0. Governor Meyner signed the bill into law on the same day as the House vote.

In Delaware the compact was viewed favorably since it made the state equal with the much bigger upstream states. The Delaware House passed the bill 32 to 0 on May 15, and the Senate approved it by a vote of 16 to 0 on the following day. Governor Carvel signed the bill into law on May 26.

It was in Pennsylvania that the last compact bill had failed. Opposition to the new bill was immediately forthcoming, particularly from large steel and oil companies, who feared charges for their water withdrawals. Governor David Lawrence, however, persuaded the industries to hold off their opposition by promising to get their concerns addressed by the federal compact legislation. (Section 15.1(b) of the Delaware River Basin Compact fulfilled Lawrence's promise.) In spite of the initial opposition, support for the compact was strong in Pennsylvania. The mayor of Easton, in fact, began lobbying to have the compact commission headquartered in his city. The compact legislation passed the Pennsylvania Senate on June 14 with a vote of 48 to 2 and the Pennsylvania House on June 22 with a 179 to 7 vote. On July 7 Governor Lawrence signed the bill. For the first time an interstate water compact for the Delaware River Basin had made it through all four states.

In Washington, President Kennedy appointed Secretary of the Interior Stewart L. Udall to be his chief decision-maker on the compact. Kennedy also ordered an investigation of the constitutionality of the federal-state partnership agreement. (The constitutionality issue had also been explored by the compact drafters. Two Columbia University law professors hired by the DRBAC concluded that the compact was legal. The Justice Department later concurred with this opinion.) Internally, various federal agencies with water-resource responsibilities opposed the compact because of fears that the new commission would infringe on their territory.

The compact legislation had been introduced in Congress in early 1961 by Representative Walter of Easton. Walter was now serving his fifteenth term in office and was highly influential in both the House and the Senate. He would play a key role in the bill's adoption. In the Senate, the compact legislation was introduced by Senator Joseph Clark, the former mayor of Philadelphia.

At first, the Kennedy administration appeared sympathetic to the compact idea. On April 13, 1961, however, Udall shocked the states when he announced his opposition to federal participation in the compact. The occasion was a meeting in Washington with the governors and other officials of the four states. The announcement brought anger and dismay. Governor Meyner was so angry that he jumped to his feet, yelling to Udall, "You've just killed it."[10] Senator Joseph Clark, also at the meeting, accused Udall of being surrounded by twenty-five "melancholy babies"—the twenty-five federal agencies involved in the water field.

In his defense, Udall raised the as yet undecided issue of constitutionality. Opponents of the compact had noted that the four states could outvote the federal government on the new commission. Shortly after the Washington meeting, a new issue was raised when Alex Radin, president of the American Public Power Association, complained to President Kennedy that the compact did not contain a preference clause for public power. Udall and President Kennedy added the preference clause to their list of what was wrong with the compact.

In spite of their initial dismay, the Delaware River Basin states did not give up. Lobbying for the legislation continued. In the Senate, the eight Delaware River Basin senators formed an informal committee to collectively push for the compact's passage. In the House, Representative Walter maneuvered the compact legislation through the House committee structure without having hearings. On Thursday evening, June 29, the House passed the measure 257 to 92 after voting down several attempts to kill it. After the House vote, Walter managed to have the legislation assigned to the Senate Judiciary Committee rather than the Interior and Insular Affairs Committee. The latter was chaired by Senator Clinton P. Anderson (D-N.M.), a good friend of Stewart Udall.

Governor David Lawrence's influence with President Kennedy was a major factor in convincing the Kennedy administration to support the compact, or at least not oppose it. Kennedy credited the governor for delivering Pennsylvania to him during the extremely close 1960 presidential election. It was a political debt of the best kind, and the various activities on behalf of the compact finally paid off. On August 15 Udall formally announced that the Kennedy administration favored the compact. Udall himself thought the interstate-federal compact was an "oddball" that should not be used in other river basins.[11]

The disputed preference-clause issue was left to the Senate to decide. In the last days of August, the preference-clause advocates, led by Udall, clashed with their opponents at the hearings of the Judiciary Committee. Senator Clark personally favored the public power preference clause but feared that the issue would kill the compact. He argued that the issue could be settled when individual dam projects came up for funding. Ultimately, this philosophy was accepted by both sides. On September 16, 1961, therefore, the Senate passed the compact legislation without a dissenting vote.

President John F. Kennedy signed the bill into law on September 27. Thirty days later, the Delaware River Basin Compact was law. In spite of his administration's lukewarmness to the compact, Kennedy hosted the official signing ceremony, held November 2, 1961, at the White House. All the governors of the four basin states except Rockefeller attended.

Six weeks later, on December 13, 1961, Secretary of the Interior Udall, Governor Rockefeller, Governor Meyner, Governor Lawrence, and a representative of Governor Carvel gathered at the Nassau Inn in Princeton, New Jersey, for the first meeting of the Delaware River Basin Commission (DRBC). Officers were chosen and temporary staff appointed until a permanent executive director could be chosen. On May 23, 1962, the Delaware River Basin Advisory Committee was officially dissolved. Its staff had been serving as the temporary staff of the new commission. Two of the DRBAC staff moved over to DRBC: W. Brinton Whitall became the commission secretary, and William Miller, the author of the compact, became its general counsel. Incodel lasted slightly longer. On January 1, 1963, it was formally absorbed by the Delaware River Basin Commission, and many of its staff also became DRBC employees.

While the Delaware River Basin Compact was being drafted and enacted into law, the U.S. Army Corps of Engineers was winding up its survey. The

Signing ceremony for the Delaware River Basin Compact on November 2, 1961, at the White House. Joining President Kennedy at the table are (*left to right*): Governors Robert B. Meyner (New Jersey), Elbert N. Carvel (Delaware), and David L. Lawrence (Pennsylvania). (White House Photo, DRBC Collection)

preliminary reservoir plan had been released in January 1959. Through 1959 and early 1960, the detailed planning volumes were released in draft form. In March 1960 a forty-page *Information Bulletin* was published for a series of public hearings held in the following two months. The Philadelphia District then finalized the report and sent it up the Corps' chain of command for approval.

Among the hundreds of pages of information on economics, population projections, demand data, costs and benefits, geological structures, water quality, and much more was a comprehensive fifty-year water-resources plan for the Delaware River and its tributaries. It was primarily dam oriented. From an initial 193 potential dam projects, the Corps had selected 19 dam projects for construction. Of these, eight were recommended for immediate authorization by Congress. All were to be designed for water supply, flood control, and recreation. They were:

1. Beltzville Reservoir, on Pohopoco Creek, four miles from the Lehigh River at Lehighton, Pennsylvania, was to contain 13 billion gallons of water-supply storage. Beltzville was recommended for construction by 1965. It was completed in February 1971.
2. Blue Marsh was planned for Tulpehocken Creek, six miles from the Schuylkill River at Reading, Pennsylvania. Blue Marsh was to be completed by 1969 and was to contain 4.5 billion gallons of water-supply storage. It was completed in April 1979.
3. Bear Creek Reservoir was an existing flood-control reservoir that was completed in 1961 in the upper Lehigh area. The Corps' plan recommended expanding the project to provide 22.8 billion gallons of water-supply storage by 1989. This dam was renamed the Francis E. Walter Dam in 1963 after Representative Walter died. Expansion of this flood-control dam has not occurred to date.
4. Trexler Dam was to be built on Jordan Creek, eight miles from Allentown, Pennsylvania, in the Lehigh River Basin. The project was to provide 7.9 billion gallons of water-supply storage by 1972. It has never been built.
5. The Prompton Project was an enlargement of a flood-control reservoir that had been completed in 1960 on the Lackawaxen River above Honesdale, Pennsylvania. The planned enlargement would have added more than 9 billion gallons of water-supply storage by 1974. The enlargement has not been made to date.
6. Aquashicola was to be another Lehigh River Basin dam, this one on Aquashicola Creek about three miles from Palmerton, Pennsylvania. The reservoir was to contain 1.3 billion gallons of storage, which was desired by 1981. It has never been built.
7. The Maiden Creek Project was to be located twelve miles north of Reading, Pennsylvania. It was scheduled to be built by 1982 with a water-supply storage of 24 billion gallons. It has never been built.
8. Tocks Island Dam was the queen of the eight projects. Located on the Delaware River six miles above the Delaware Water Gap, Tocks represented over 58 percent of the intended storage capacity of the Corps' reservoir plan. Unlike the other seven projects, hydroelectric power was to be generated at Tocks. Construction costs were estimated to be $146 to $177 million, and completion was set for 1975. None of the other dam projects approached the magnitude of the Tocks Island Dam project.

In addition to the eight projects recommended for immediate federal authorization were three other recommended projects. These were to be built

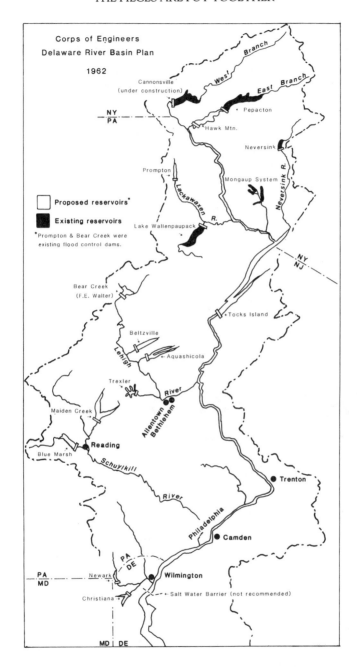

Corps of Engineers
Delaware River Basin Plan

1962

Cannonsville
(under construction)

West Branch

East Branch

Pepacton

NY
PA

Hawk Mtn.

Neversink

Prompton

Mongaup System

Neversink R.

☐ Proposed reservoirs*

■ Existing reservoirs

Lackawaxen R.

Lake Wallenpaupack

*Prompton & Bear Creek were
existing flood control dams.

NY
NJ

Bear Creek
(F.E. Walter)

Tocks Island

Beltzville

Lehigh

Aquashicola

Trexler

River

Maiden Creek

Allentown
Bethlehem

Reading

Blue Marsh

Schuylkill

Trenton

River

Philadelphia

Camden

PA
DE

PA
MD

Newark

Wilmington

Christiana

Salt Water Barrier (not recommended)

MD | DE

for water supply and recreation. Since they contained no flood-control benefits, the projects were not eligible for federal construction. The three were:

1. The Hawk Mountain Project was to be built in New York State on the
   East Branch, Delaware River, downstream from Pepacton Reservoir. It
   was scheduled for construction by 2001 with a water-supply storage of 95.5
   billion gallons. Hawk Mountain was the second-largest dam in the Corps
   plan and the only one besides Tocks Island to contain hydroelectric-power
   generation. It has never been built.
2. The Newark Project was to be located on White Clay Creek near the
   Delaware and Pennsylvania border. With 10 billion gallons of storage, it
   was the smallest of the recommended reservoirs. The dam was to be built
   by 1975, but it never was. In the mid-1980s, a bi-state park was created
   using land acquired for the reservoir.
3. The Christiana Project was located near the Newark Project, but on the
   Christiana River in Delaware. It was to contain 12 billion gallons of
   water-supply storage. Scheduled for 1980, it was never built.

The Corps' 1960 Delaware River Basin reservoir plan was distinctly different
from the Incodel Integrated Water Project that had preceded it. Cannonsville
Reservoir, an intrinsic component of the Incodel plan, was under construction
as the Corps prepared its plan and was thus a plan "given." The two Delaware
River Basin plans, however, delivered similar minimum flows to the Delaware
River. Under the full development of the two-phased Incodel plan, maintenance of a minimum flow at Trenton of 4,800 cfs would have been possible.
The construction of the Corps' reservoir plan yielded an equivalent flow of
4,720 cfs. Water diversions to New York and New Jersey, which had been
defined in the Incodel plan, were largely undefined by the Corps. By comparison, the three New York City reservoirs, when completed, were thought to be
capable of delivering a minimum Trenton flow of 2,840 cfs. The drought of the
early 1960s, which began shortly after the Corps' work was completed, reduced
the safe yields of all existing and proposed reservoirs, however.

A main-stem dam that failed to make the recommended list was the "Salt
Water Barrier" in the Delaware Estuary. The Salt Water Barrier idea was
largely promoted by Delaware officials, with some interest being shown by
one or more New Jersey persons.[12] Basically, the Salt Water Barrier would
have been a 53,300-foot-long, 30-foot-high dam across the tidal river near
New Castle, Delaware. There were to be navigation locks to allow ships to
pass. As its name suggests, the purpose of the Salt Water Barrier was to keep
saltwater from intruding north of Wilmington, thus creating a freshwater lake
between Wilmington, Delaware, and Trenton. The idea was not a compo-

nent of the Corps' original plan of study, but in 1958 proponents of the dam got the Senate Public Works Committee to authorize the study.

The Corps held hearings on the Salt Water Barrier, conducted engineering feasibility studies, and even developed a physical model of the dam at its Vicksburg Waterways Experiment Station. The project was opposed by shipping and commercial interests, various estuary cities, and the oyster industry. The other basin dam supporters gave lukewarm support for the study since they feared that the project would overshadow the Tocks Island Dam project. There were serious questions, too, about all the pollutants that would be dumped into the freshwater lake. The Salt Water Barrier was found to be technically feasible but uneconomical.

On March 28, 1962, the new Delaware River Basin Commission met for the second time. At this meeting, the commission adopted its *First Phase Comprehensive Plan*. The plan had been hastily pulled together by DRBC's temporary staff. Included in the first plan were the dam projects recommended in the *Delaware River Basin Report*. The Corps, on the other hand, had yet to finish its own review and approval process.

The Corps of Engineers completed its review of the *Delaware River Basin Report* in April 1962. In August Congress received the report and ordered it reprinted as House Document 522. In the fall congressional hearings were held, but they were brief, with only the Corps of Engineers testifying.[13] On October 23, 1962, Congress passed the Flood Control Act of 1962 (P.L. 87-874). The act, passed at the last minute before Congress took its preelection recess, contained nearly two hundred public-works projects nationwide. Section 203 of the act authorized the "project for the comprehensive development of the Delaware River Basin" as contained in the Corps of Engineers' report. The authorization contained a total price of $192.4 million.

By the end of 1962, proponents of a Delaware River dam had every reason to rejoice. All the elements needed to build Tocks Island Dam had been accomplished: federal authorization, the creation of the long-sought interstate agency, and the creation of general public support for the project. None of these had been accidents. The creation of the Tocks Island Dam project had been a well-orchestrated endeavor. After this accomplishment, building a dam would seem easy.

7

In All Its Glory

The easiest way to find Tocks Island is by canoe. The island can be seen
through the trees from the Old Mine Road on the New Jersey side of the
Delaware, but the narrow road requires all of a driver's attention. The island
is just downstream from Smithfield Beach, an important recreation spot in
Pennsylvania, and exactly 6.4 miles upstream from the Delaware Water Gap.
Too small for much else, Tocks Island's only purpose has been to provide a
haven for wildlife and canoeists. Since 1954 Tocks Island has been owned by
the State of New Jersey; the Worthington family owned it before that.

The Tocks Island Dam contained in the *Delaware River Basin Report* was
located across the northern (upstream) tip of Tocks Island. Various geological
studies conducted by the Corps of Engineers had ruled out the possibility of a
concrete dam at the site, but the Corps' 1957 study had indicated that an
earth-and-rock dam might be feasible. The *Delaware River Basin Report*, therefore, outlined a steep-sided, earth-and-rock dam running 3,200 feet across the
narrow Minisink Valley. The planned height of Tocks Island Dam was about
160 feet above the Delaware River bed, and its width varied from 400 to 900
feet. The narrow valley enclosed by mountains was the perfect place to mix
spectacular natural scenery with an impressive engineering structure.

The planned dam spillway was on the New Jersey side of the Delaware. The
1957 preliminary Tocks report had placed this spillway on the Pennsylvania
side, but the Corps' subsequent investigations had found geological problems
at the site. On top of the spillway, ten gates, each forty feet long and thirty-

five feet tall, were planned. These would regulate the reservoir releases. Releases were also to be made through twin twenty-two-foot-diameter conduits passing through the dam structure. The conduits were to deliver water to the dam's powerhouse. The proposed plant consisted of two 33,000-horsepower turbines connected to 23,000-kilowatt generators and other necessary equipment. The power plant, like the dam spillway, was to be located on a ledge blasted out of the side of Kittatinny Mountain.

Tocks Island Dam was much more than power plants and dam structures. It was water. Tocks Island Reservoir was to be a narrow thirty-seven-mile-long lake running up the U-shaped valley to Port Jervis. A nine-mile arm would run up the Flat Brook Valley in New Jersey and a smaller one up the Bushkill Valley in Pennsylvania. The reservoir was to have a total surface area of 12,300 acres, with depths up to 140 feet. Behind the dam would be almost 250 billion gallons of water. The exact amount of water behind Tocks Island Dam, however, would vary seasonally and even daily.

The Tocks Island Project was a true multipurpose project. The 1962 authorization had required Tocks to provide water supply, flood control, hydroelectric-power generation, and recreation.[1] The reservoir's storage capacity was allocated accordingly. The allocation began with an inactive long-term storage pool from the reservoir bottom to an elevation 356 feet above sea level. The inactive pool contained about 26 billion gallons of water. From elevation 356 feet to elevation 410 feet was the planned active pool, to be used for water supply, power generation, and recreation. The active pool was to contain about 134 billion gallons of water. Drawdowns of the active pool could add 980 cfs to the normal flow of the Delaware on a sustained basis. This water, more than 600 million gallons per day, was the water that the planners and engineers in the lower Delaware River Valley had been talking about for years.

The flood-control component of the reservoir was to be above the active pool to an elevation of 428 feet. The flood-control pool would normally be empty except when filled by runoff from rainstorms or snowmelt. The surface elevation of the flood-control pool was set at elevation 428 feet, the highest elevation that could be used without building dikes around Port Jervis. In most cases the planned flood-control capacity was sufficient to prevent all the potential flood damage downstream from the dam. During a really big flood, one similar to that of 1955, Tocks Island Dam, along with the dams on the Lehigh and Lackawaxen rivers, could significantly reduce the flood stage at Trenton and elsewhere.

The scenic attributes of the Minisink Valley and the nearness of millions of potential visitors made recreation a significant feature of the project, too. At first it was estimated that almost seven million people would visit the reservoir and surrounding lands each year. To accommodate these visitors, the Corps' planners envisioned the development of ten recreation areas between Milford and the Delaware Water Gap. Recreation was also tremendously important to the dam project's economic feasibility. Since it was to use water already in the active pool, recreation was a relatively cheap addition, yet tremendous dollar benefits could be calculated. Recreation was also the most exciting feature of the Tocks Island Project and the easiest to sell to the general public.

Hydropower was a side benefit, too, since it also used water dedicated for some other purpose. The planned power capacity of Tocks was greater than that of Lake Wallenpaupack, the largest hydroelectric-power operation in the Delaware River Basin. On an annual basis, the plant at Tocks was to have an estimated dependable production of 281.5 million kilowatt-hours.

Tocks Island Dam was not free. The original price tag on the dam package was $122.4 million. This figure broke down to $14.2 million for flood control, $58.2 million for recreation, $21.5 million for power facilities, and $28.5 million for water supply. The U.S. government was assigned all the expenses except the water-supply costs, which were to be paid back by users through a local sponsor, the Delaware River Basin Commission.

The Tocks Island Dam Project outlined in the *Delaware River Basin Report* was far from the last word. Following project authorization, the next step was to conduct preconstruction planning. This phase of the process consisted of detailed studies, or design memoranda, on each part of the project. Twenty-two design memoranda were initially planned for the project. These were to be followed in turn by the detailed engineering designs that would be used to build the project.

The most important design memorandum in the preconstruction planning process was the general design memorandum, approval of which established the major features of the project: reservoir capacity, reservoir water-storage allocations, the location of the dam axis, and the various engineering details of construction. Firm costs and benefits were also to be presented in order to reaffirm the original project justification.

The original Tocks schedule called for construction to begin in the fall of 1967.[2] Land on the New Jersey side of the dam site would be purchased and the first construction contracts awarded. Eight construction seasons later, the dam would be completed. It was envisioned that by 1971 Tocks Island Dam would be far enough along to hold back flooding if it occurred. In the following year, the reservoir would begin filling, and the whole project would be fully operational sometime in 1975.

Through most of the preconstruction planning, the most worrisome technical problem was where exactly to put Tocks Island Dam. The Minisink Valley from Wallpack Bend to the Delaware Water Gap is not a good place to build dams. This had been known ever since the New York District had first studied and rejected the Tocks Island site back in 1942. In its 1957 study, however, the Philadelphia District had concluded that an earthen dam structure was feasible at the site. By the early 1960s, too much was at stake to think otherwise!

The search for the exact site for Tocks Island Dam began in 1964. In the 1930s and 1950s, the Corps of Engineers had made exploratory borings of the foundation conditions in the region, but many more tests were needed. Ultimately, the Corps removed a total of seven miles of core samples in the Tocks Island region. These were augmented by twenty or more miles of seismic surveys. The core samples were shipped to the Corps' testing laboratory, where a variety of special engineering tests were conducted.

Finding suitable foundation conditions for Tocks Island Dam would not be easy. Because of the unusual site problems, a special Board of Consultants was hired to help the Corps. The three men on the board were experts in their fields: Dr. Arthur Casagrande, a Harvard University professor and expert on earth-and-rock dam engineering; Francis B. Slichter, a consulting engineer with worldwide experience in dam design; and Dr. S. S. Philbrick, a Cornell University geologist with extensive dam-site selection experience.

The site problems at Tocks Island had been created by the ice ages. The ice had wrecked the geology of the area and filled the ancient valley with a variety of materials up to 200 feet deep. Four major types of materials had been left for the dam planners. Till material, which had been laid down by melting ice, consisted of coarse to boulder-size rocks and fine-grained rock flour generally located on top of bedrock in depths averaging 40 feet. Above the till material there was often drift material derived from the outwash from melting ice blocks. Drift material varied in size and was contained in deposits up to 110 feet deep. Lake material, found in deposits from a few feet thick to 150 feet thick, was made up of fine-grained sediments and rock flour that had settled in temporary lakes created by isolated, melting ice blocks during times when the ice temporarily dammed stream courses. Capping all the ice-derived materials were sand and gravel beds deposited by the river in the years since the ice ages. At any one location the mix of till, drift, lake, and alluvial deposits was extremely varied. Each represented a special problem for dam building.

The most critical problem facing the engineers was the scattered lake deposits. These were found not only in the valley but also under the bluff on the Pennsylvania side of the dam site. The lake deposits were sensitive to dynamic loading and, under stress, could lose strength and even liquify. Core samples containing lake deposits required special handling just to get them to the Corps' New England soil-testing lab.[3] In spite of the special handling and packaging, some of the samples would slump and liquify before reaching the lab. The extreme sensitivity of the material meant that earthquakes could cause a dam failure. The deposits under the Pennsylvania bluff were particularly worrisome since the rise and fall of the reservoir pool could also affect their stability. There were other serious geological problems facing the engineers, but the lake deposits were the trickiest by far.

The search for the final site for Tocks Island Dam wound up its first phase in December 1964. The Board of Consultants advised the Corps that the dam site presented in the *Delaware River Basin Report* was *not* feasible.[4] In fact, a dam could not be built in most of the lower Minisink Valley. The only hope for a suitable dam site was in a 3,000-foot zone running from the southern (downstream) tip of Tocks Island to Labar Island. This area was then selected for further study.

The initial studies had also indicated that seepage would be a special problem in the valley and that an extensive seepage-control program would be needed. Seepage, if not controlled, could affect the dam's stability. Assuming that a feasible site could be found for the dam, seepage control would add considerably to the project costs.

Although the studies continued, by the end of 1966 a site for Tocks Island Dam had been found.[5] Instead of crossing the valley at the northern tip of Tocks Island, the dam would be located about a hundred feet below the southern tip of the island. The new site contained quantities of coarse material that could, it was hoped, be used to stabilize the pockets of lake deposits. The new site was far from ideal, but the Board of Consultants believed it could work.

The geology problems also changed the design of the dam structure. The dam, as originally proposed, was to be steep-sloped, with slopes of 3 to 1 and 2.5 to 1. Construction of a steep-sloped dam, however, required the removal of the lake deposits. This was considered feasible at the new site, but it was a complex, costly endeavor with various unknowns. To get around this problem it was decided to build a flat-sloped dam with a width 20 times its height and an average slope of 10 to 1. The flat-sloped dam was to extend 1,600 feet upstream and downstream from the dam axis. This allowed some of the fine-grained deposits to remain since the weight of the dam was spread over a much wider area.

While the Corps of Engineers was searching for the final dam site, it was also conducting another study that would influence the final configuration of the Tocks Island Dam Project. By the mid-1960s, the Delaware River Basin was in the midst of an extremely serious drought. Reservoir levels throughout the region declined drastically, and flows in the Delaware River reached all-time lows. A crisis was precipitated on June 14, 1965, when New York City ceased making its Montague flow releases. The drought had caught the city off-guard, forcing it to take the drastic action.

The city's failure to live up to the Supreme Court Decree panicked the downstream states. As the water situation began to unravel, the new Delaware River Basin Commission began trying to resolve the crisis. Through lengthy and sometimes heated negotiations, solutions were found that avoided another Supreme Court fight. Available water supplies were stretched as far as possible, and the basin managed to limp through the emergency. The drought of the 1960s would have profound ramifications for water management in the Delaware River Basin.

Among the offshoots of the 1960s drought was a study of the benefits of raising the Tocks Island Reservoir pool level from its original 428-foot elevation to as high as 450 feet. The extra water would be used for repelling saltwater intrusion in the Delaware Estuary during severe droughts. The potential increase in the size of the reservoir generated considerable anger in the Tocks Island region. The higher pool level meant that high levees would have to be built at Port Jervis and Matamoras.

The pool-level study was not a technical study but an economic study to determine which level resulted in the maximum benefits at the smallest unit cost. The study found several points of maximization. For water supply, power, and recreation, the maximum benefits occurred at elevation 409 feet. As the pool level rose above this elevation, recreation benefits declined because the flat shore areas needed for recreational facilities were flooded. Flood-control benefits, normally dry storage that did not affect recreation, rose as the pool level was raised. The ultimate point of maximization occurred at elevation 436 feet, the maximum practical height of levees in the Port Jervis area.

Significantly, the drought considerations that had triggered the study reduced the benefits of Tocks Island Dam. In other words, water stored to retard saltwater intrusion in the lower Delaware River Valley could not be economically justified when compared to the other benefits of reservoir storage. In less than ten years, however, this philosophy would change. Water-supply storage would come to mean water for salinity control, something distinctly different from the water-supply storage considered in the pool-level study.

As a result of the pool-level study and the selection of a different dam site, the size of Tocks Island Reservoir increased over 9 percent to a capacity of 275.4 billion gallons. The reservoir behind the dam was thus reallocated, with the active water-storage pool increasing about 4 percent and the flood-control pool about 15 percent. The allocation of the reservoir's water was important because it determined the dollar benefits of the project. These had to be greater than the project's costs or else the dam could not be built.

Tocks Island Dam, however, had help in its justification. The project was one of the first large-scale federal dam projects to use recreation benefits in its benefit-cost calculations. By 1975 up to 44 percent of the project's benefits was being assigned to recreation. This aided the benefit-cost ratio tremendously since recreation facilities are cheap additions to a dam project; the same dam and reservoir are built whether recreation is an authorized purpose or not. A simple calculation shows that any dam with a benefit-cost ratio below 1.8 and recreation benefits totaling 44 percent of the total project benefits has a benefit-cost ratio less than 1.0 if recreation is not used in the calculation. Recreation, more than any other project purpose, made Tocks Island eligible for federal funding. Perhaps this is what past studies of a Delaware River dam had told its sponsors. A main-stem dam did not make economic sense from a purely water-supply or flood-control viewpoint.

Tocks Island water was not to be owned solely by the federal government. The Water Supply Act of 1958 (P. L. 85-500) allowed water-supply storage to be added to federal reservoir projects *if* a nonfederal sponsor repaid the associated costs. The Delaware River Basin Commission became the nonfederal sponsor when it agreed in September 1965 to buy the water-supply storage in Tocks Island Reservoir.[6] This action of the commission made it the Corps' partner in the Tocks Island Dam Project.

The dam that emerged from the preconstruction planning was vastly different from the dam described in the *Delaware River Basin Report*. The biggest changes were, of course, the dam's new location, the reservoir's larger size, and the dam's shape. There were other changes, too. The spillway, intakes, powerhouse, and related structures were reversed in position, and the ten spillway gates were replaced by six larger gates. These controlled flood levels by containing and then releasing the large quantities of water from rain and snowmelt runoff. Normal reservoir releases were to be made through twin tunnels, each about 1,300 feet long and 24 feet high. These were to be bored into the side of Kittatinny Mountain. The landward tunnel would route water to the powerhouse, but the riverward tunnel could also deliver water directly to the Delaware River. A larger powerhouse had been found feasible, and the potential generating capacity was increased to 70,000 kilowatts, or 50 percent more than originally planned.

The flatter dam design allowed the Tocks developers to plan a parklike setting at the dam. The focus of this area was a permanent visitors center located on top of the Pennsylvania dam abutment. From the visitors center, a network of guided and self-guided tours of the dam was planned. Below the visitors center, on the gently sloping dam embankment, would be walkways and landscaped picnic areas. For some people, there could be beauty in a dam:

> Tocks Island Dam will stand as a strong physical demarcation between the mountain lake landscape upstream and the pastoral landscape downstream. In the proposed plan for its architecture and develop-ment, Tocks Island Dam will serve as a transition buffer between the two landscapes, the intent being to make the dam an intrinsic ele-ment of the new natural scene, just as the farmhouse, barn, silo and other works of man became part of the valley's agricultural scene. This intent will be realized in a naturalistic park and through architectural design that deemphasize the purely mechanical functions of the dam structure, bringing them into scale and harmony with the natural setting.[7]

The "naturalistic park" and Tocks Island Dam were pipe dreams. By all rights a host of dignitaries should have descended on the Minisink in 1967 and, grasping ribbon-bedecked shovels, turned over the first soil. The cere-mony did not take place, however. Try as they might, the Corps of Engineers could not get construction started in the turbulent decade of the 1960s.

It was cold, hard cash, or the lack of it rather, that kept dam building from occurring. Tocks Island Dam had a tremendous cash-flow problem because of its size, its questionable cost estimates, and the geological problems of the dam site. This was an internal problem that could be laid at the feet of the Corps of Engineers. Externally, the project was competing for dollars that were being sent to fight Communism in Southeast Asia. During the second half of the 1960s, there just wasn't enough money in the federal treasury to build grand dam projects in the Minisink Valley and also wage war.

Tocks Island Dam was in financial trouble immediately after Congress authorized the project in 1962. Many persons would claim that the Corps had deliberately underestimated the costs of the dam to ensure congressional authorization of the project.[8] One thing was sure: The cost estimates for Tocks Island Dam were never static. As originally authorized in 1962, Tocks Island Dam carried a price tag of more than $90 million. By the time Con-gress appropriated the first funds for the project, the price had risen to $95 million due to inflation. As planning got under way, however, the cost estimates escalated dramatically.

The changes in dam design due to the geological problems added almost $16 million to the cost estimate between July 1, 1964, and July 1, 1965. The need for protective works in the upper part of the reservoir pool added

Architect's model of Tocks Island Dam and visitor facilities. The model depicts the second, or final, dam design. (DRBC Collection)

another $14 million. Inadequate estimates of the amount of cemeteries, high-ways, schools, and utility lines that needed relocation added another $12 million to the estimate by July 1966. Land acquisition cost estimates in-creased by $15 million, and another $8 million was added for other land-related costs, such as the lands for wildlife migration measures. As each day went by, seemingly more and more money was needed for the project.

The net effect of all the cost increases was truly incredible. By July 1967 the estimated cost of Tocks Island Dam was placed at $198 million, a $100 million jump in less than three years. Congress, which had seen a $140 million price tag in the previous year and a $98 million price before that, was jolted. In the House, the Public Works Committee ordered a staff study of the project's benefit-cost ratio. The ratio had to be greater than one for the project to survive.

As 1967 closed, *Time* magazine published an editorial entitled "How to Cut the U.S. Budget."[9] The editorial reflected President Johnson's recent call for cuts in the national budget. *Time* noted that Congress had a $4.6 billion public works bill in fiscal year 1968 and that "the rich aroma of pork converts even the most ardent budget cutters into big spenders." High on the magazine's hit list was the "Delaware River—Tocks Island Park," which it considered nonessential.

The most immediate impact of the fast-rising costs was the slowdown of the project. As a consequence of the severe mid-1960s drought, President Johnson ordered the Corps to accelerate the construction of Delaware River Basin dams. Congress appropriated enough money to speed up preconstruction planning by one year, but, soon afterward, the Corps announced that the extra money would barely keep the project on its original schedule. Tocks Island Dam was beginning to get a reputation as a bottomless pit for federal appropriations.

The staff study ordered by the Public Works Committee in August 1967 was prepared by the Federal Bureau of Investigation. The report was never made public, but its contents were leaked to the press in early 1968.[10] The investigators decided that Tocks Island's benefit-cost ratio had fallen so low that its feasibility was questionable. The estimated costs had now risen to $203 million, and the project had a benefit-cost ratio of approximately 1.5 or 1.4, according to the Corps. However, the U.S. Fish and Wildlife Service was now claiming that the dam's operations would destroy as much as two-thirds of the Delaware Bay oyster industry, resulting in damage of more than $4 million. The House investigators believed that oyster losses over $4.225 million would drop the benefit-cost ratio below one. The Corps of Engineers, of course, disputed these findings.

Tocks Island Dam survived the Public Works Committee report. In 1968, however, the project came under criticism in the Senate. During hearings in March 1968, Senator Allen J. Ellender (D-La.), chairman of the Subcommittee on Public Works, publicly questioned the costs of the dam and warned that the costs and other issues surrounding the project had better be resolved. In the House, Tocks Island Dam received similar criticism from the Public Works chairman, Michael Kirwan (D-Ohio). Even Congress, which usually favored expensive pork-barrel projects, was beginning to wonder if the country could afford a costly project like Tocks.

In 1969 the estimate for the Tocks Island Project was revised upwards to $214 million, an $11 million increase from the previous year. The price increase was significant since in the first six years of the project Congress had appropriated only $9.34 million for the project. Criticism of the project became harsher. Senator John Sherman Cooper (R-Ky.), in particular, became a new critic of the dam project. Cooper was an influential member of the Public Works Committee.

In 1969 funding for the dam was delayed for the first time. Senator Ellender requested that the General Accounting Office investigate the cost allocations, dam-site selection, and the Corps' land-acquisition process. The GAO subsequently raised several questions about the allocation of costs between water supply and recreation. It believed that the Corps' water-supply benefits were understated by $21 million and that the recreation benefits were overstated by $8 million. This meant that water supply accounted for more than 30 percent of the project costs. The Water Supply Act of 1958 required the Corps to have contracts for the sale of water in excess of 30 percent before construction could begin. The Corps argued that water supply was only 29.7 percent of the project costs and that a contract was not yet needed. It and DRBC were still negotiating the latter's purchase of water storage.

The GAO investigation was primarily a bookkeeping effort, although it did demonstrate that the costs and benefits of a project like Tocks Island Dam were highly subject to manipulation. Ellender's reaction to the report was to insist that the Senate Appropriation Committee report contain implicit criticism of the way Tocks Island benefits were calculated. He wanted to make sure that a precedent was not set for future water projects. After 1970, though, the cost issue was replaced by a concern for the dam's environmental impacts. The cost estimates kept climbing, however. By 1975 the price tag on Tocks Island Dam was estimated to be $400 million and climbing.[11]

Tocks Island's spiraling cost estimates meant that each dollar appropriated by Congress did less. To stay on schedule, higher and higher appropriations were required. Appropriations for the dam, however, never really appeared in large enough amounts because of the Vietnam War. The delay in funding brought about by the war was the single most important reason the dam was never built.

The idea of a dam across the Delaware River and the efforts to make Vietnam a Communist country originated in the early 1920s and took thirty years to evolve. The mid-1950s found the federal government increasing its involvement in both the Delaware and Vietnam. The flood of 1955 had caused the intervention in the Delaware. In Vietnam, the defeat of the U.S.-financed French army at Dien Bien Phu in 1954 had led to the introduction of American advisors into South Vietnam. The first American casualties occurred in 1959, the same year that drafting of the Delaware River Basin Compact began. In the two years between the publication of the *Delaware*

River Basin Report and the authorization of the Tocks Island Dam Project, the number of American troops in South Vietnam increased from 1,000 to 11,000.[12] Three years later, American troops in South Vietnam numbered almost 150,000, and bombing raids over North Vietnam were averaging 4,000 per month. Tocks Island Dam, meanwhile, was getting a full year of congressional appropriations for the first time.

The impact of the Vietnam War on Tocks funding was almost immediate. Early in the war effort, President Johnson and his advisors decided to fight the war without raising taxes. Just as the dam was being funded, domestic spending programs were being cut. Inflation, heated by the budget deficit, began rising, making the price of everything higher. During the life of the Tocks Island Dam Project, the Vietnam War cost somewhere between $141.3 billion[13] and $171.5 billion.[14] A war costing nearly $32,000 per minute did not leave much room for expensive dam projects.

The war hit home in early 1966 when the federal budget for fiscal year 1967 was proposed. In his budget message, President Johnson cautioned, "even a prosperous nation can not meet all its goals all at once. For this reason, the rate of advance in new programs has been held below what might have been proposed in less troubled times."[15] The Corps of Engineers received about $1.2 million in the fiscal year for Tocks Island. The original construction schedule had called for the start of construction in fiscal year 1967, but the design work and land acquisition were way behind schedule.

In early 1967 the Vietnam War was costing more than ever. Large federal cutbacks in domestic programs loomed on the horizon. President Johnson's goal was to cut ongoing programs by deferring them three to six months.[16] As a result, the Corps of Engineers kept its budget requests low. Corps headquarters, for example, requested only $3 million for Tocks Island Dam, even though the Philadelphia District had requested $5.65 million. Congress appropriated $4 million for the project.

In 1968 the Vietnam War became the longest war ever fought by the United States. By the middle of the year, troop strength in South Vietnam reached 534,700. President Johnson called for "sacrifices and hard choices"[17] and a "determined effort to slow the pace of federally financed construction programs as much as possible."[18] In case Delaware River Basin dam boosters did not get the message, the president stated, "I am recommending that ongoing water resources projects be continued at minimum rates. In many cases this will require a delay in present construction schedules."[19]

As the result of the call for spending cuts, the Corps of Engineers drastically reduced its fiscal year 1969 budget request and got even less than it requested. The Philadelphia District sent up a request for $9.2 million, which reflected the approaching start of dam construction. The Corps' Washington office reduced the district's request to $5 million, the Budget Bureau reduced it to $4 million, and Congress appropriated $3.88 million. Finally, as the result of the Revenue and Expenditure Control Act of 1968, $1.83 million of the final appropriation was cut. This left the Philadelphia District with 22 percent of its original budget request. With Tocks Island Dam costs continuing to spiral, it was easy to see that the project was going nowhere.

In the following fiscal year, the Philadelphia District did a little better, getting a $4 million appropriation for Tocks Island Dam. With the war winding down, the public-works funding crisis was easing, although the Nixon administration did cut public-works funding by $142 million nationwide. Each succeeding fiscal year would be better than the preceding one, with fiscal year 1970 being the last year that Vietnam War spending significantly affected Tocks funding.

The damage had already been done, however. The original schedule for the Tocks Island Dam Project had assumed that the dam would be nearing completion by 1971 and would be capable of holding back a flood if one occurred. Because of the war and the dramatic increases in the project costs, the Corps had purchased only 7 percent of the land needed for the dam project and had issued no construction contracts.[20] Tocks Island supporters were now pushing for funding levels that would accelerate dam construction once it began.

Because of the delays, time caught up with Tocks Island Dam. On January 1, 1970, President Nixon had signed into law the National Environmental Policy Act of 1969 (P.L. 91-190). Section 102(c) of the act required federal agencies to prepare environmental impact statements. These studies were to address in detail each project's adverse impacts on the environment. Had Tocks Island Dam been under construction, it may have been exempted from the full scrutiny of the Section 102(c) process.[21] This process would usher in a new era for Tocks Island Dam. The dam would not survive.

8

A Central Park for Megalopolis

Recreation, more than any other feature of the Tocks Island Project, was used to sell the project to the public. The features and setting of Tocks Island Dam practically sold themselves. Picture a forty-mile-long lake nestled between wooded mountains only a half-dozen miles above the world-famous Delaware Water Gap. Streams with wild beauty tumble from the ridges into the Delaware River. Old oaks, dark groves of hemlocks, mountain laurel, dewy ponds, rhododendron, silver waterfalls, and much more speak of the beauty of the area. Reservoir recreation on the Delaware River capitalized on the historic recreational economy of the region and added a man-made feature to the area's natural amenities.

The National Park Service was responsible for the recreation section of the *Delaware River Basin Report.* In 1957 Peter DeGelleke, a resident of Warren County, New Jersey, was hired by the National Park Service to develop the recreation plan. Living near the Tocks Island region, DeGelleke was well acquainted with the scenic quality of the area. The recreational potential of the Tocks Island Dam Project excited him. After a year of study, he recommended that the federal government assume responsibility for recreation at Tocks Island.[1] Typically, state agencies developed and operated the recreational facilities at Corps dams.

The idea of a national park surrounding Tocks Island Reservoir appealed to the National Park Service. Most of its parks are unique natural areas located in the less populated West. In the East, the National Park Service had been largely limited to servicing nationally important historical areas, such as Independence Hall in Philadelphia. The lack of park-service involvement in the East was the result of land passing into private ownership long before anybody had conceived of recreation as a governmental responsibility. Tocks Island Dam thus presented an opportunity for the park service to develop a major recreational facility in the heart of the populous eastern states.

In order to qualify for federal development, the recreation features of Tocks Island Dam had to have broad regional or national significance.[2] The National Park Service concluded that this was the case and that Tocks Island Reservoir could become "the most significant non-urban recreation area in the Eastern United States."[3] This translated to millions of people visiting Tocks Island Reservoir each year. The result was that the "Tocks Island Reservoir Recreation Area" was recommended for full development by the federal government.[4]

Even as the National Park Service was putting the finishing touches on its portion of the Corps' basin report, the publicity began. In September 1959 the Water Research Foundation for the Delaware River Basin released a slick twenty-six-page pamphlet entitled *Water for Recreation—Today and Tomorrow*. The pamphlet was widely distributed by the WRA/DRB via its mailing list of twelve thousand.[5] The pamphlet was the first to call public attention to the potential benefits of the Tocks Island Dam Project.

> The waters of the lake would be excellent for such activities as bathing, fishing, and boating. Its banks could be developed in such a fashion as to satisfy the broadest range of outdoor recreational interests, with picnic grounds, sites for group camping, cabins and cottages, trails flanked by rare and attractive phenomena, scenic drives, and parking areas overlooking the water and surrounding land formations.[6]

The 1959 pamphlet attempted to show that recreation at Tocks was "clothed in a sense with National interest."[7] In the following year, the WRA/DRB got more to the point in *Tocks Island: A National Recreation Area?* The brochure called for a united lobbying effort to ensure that Congress acted on the idea for a national recreation area. In subsequent years the WRA/DRB would take credit as the first organization to publicly discuss and promote this idea.[8]

The WRA/DRB publicity campaign for the Tocks Island national recreation area continued unabated during the debates on the Delaware River Basin Compact. Speeches by WRA/DRB representatives were given, WRA/DRB-sponsored conferences were held, and material was fed to the news media. In 1962 the WRA/DRB produced a documentary film concerning the many aspects of the Tocks project, and in early 1964, a ten-page pamphlet was produced and widely distributed. Entitled *Tocks Island and Outdoor Recreation for the Crowded East,* the document glorified the potential recreation opportunities and urged readers to write Congress.

Although the average citizen remained rather oblivious to the recreation-area proposal,[9] a second audience was reached by the publicity campaign. This group consisted of special-purpose organizations with a direct interest in recreation: the vacation-resort industry, labor groups, local governments, builders, and economic-development organizations. In addition, the National Wildlife Federation, the League of Women Voters, sportsmen's clubs, and others were also recognized recreation-area boosters.

Not surprisingly, major support for the recreation area came from the Delaware River Basin Commission. DRBC was the first governmental agency outside the federal government to officially endorse the proposal. On March 28, 1962, DRBC included the recreation area in its first comprehensive plan. This action lent considerable support for the recreation area. DRBC also assisted the publicity efforts to the extent that its staff resources allowed. In its first annual report (1963) and in subsequent publications, the commission promoted the idea for a national recreation area as an intrinsic part of the Tocks Island Dam Project.

The final decision concerning the recreation area belonged to Congress. The first bill to create the Tocks Island National Recreation Area was introduced into the House of Representatives on June 21, 1962, by Representative Francis E. Walter. Companion legislation was introduced in the Senate on July 11 by Senator Joseph S. Clark and other senators from the Delaware River Basin states. The bills were sent to the House and Senate Committees on Interior and Insular Affairs. Although the Senate committee approved its bill, both bills died before enactment. Similar bills failed to be enacted in 1963 and 1964 as well.

The problem with the Tocks Island National Recreation Area bills lay in the House and Senate Committees on Interior and Insular Affairs. Both committees were attempting to get legislation passed that would create a funding mechanism for national recreation areas and parks. The Land and Water Conservation Fund Bill was stalled in the House, and Wayne Aspinall, the chairman of the Interior Committee, was not letting any new park proposals through Congress until it passed.

In the fall of 1964, the Land and Water Conservation Fund legislation (P.L. 88-578) was finally enacted into law, paving the way for the enactment of the Tocks Island National Recreation Area legislation. The final push came in 1965. In order to increase the pressure on Congress, still another pamphlet was published. This one, *Tocks Island National Recreation Area—A*

Proposal, was published by the National Park Service in order to increase the pamphlet's prestige. The WRA/DRB, however, had a heavy hand in its development, including underwriting the publishing costs.[10] The WRA/DRB then distributed 25,000 copies of the 40,000-copy pressrun.

The twenty-eight-page pamphlet, replete with full-page photographs, glorified the recreation-area proposal to its fullest. Besides describing the many amenities of the Minisink Valley, the report warned that second-home developments and subdivisions were encroaching on the area.

It concluded:

> It is difficult to conceive of such a large mountain and valley area overrun with developments and no longer an attractive amenity. Only Federal ownership of a large area—as recommended in this proposal—can prevent the development of this resource for a few. Only Federal intervention can reserve the natural scenic character for public recreation use.[11]

Tocks Island National Recreation Area legislation was introduced in early 1965 by Representative John P. Saylor of Pennsylvania, the ranking minority member of the House Interior Committee. At the same time, companion legislation was introduced in the Senate by Clark and the rest of the Delaware River Basin senators. Both the House and Senate bills were introduced in early January in order to receive low numbers. The low numbers, H.R. 89 and S. 36, were believed to have a psychological impact on passage. The numbers also assured the timely submittal of committee reports from the Department of the Interior, the U.S. Army, and the Delaware River Basin Commission.

President Johnson endorsed the recreation area on January 8, 1965. On January 22 the Delaware River Basin Commission submitted its report to Congress (acting in this regard as a federal agency). This was followed by the Army's report on February 24 and the Interior Department's report two days later. The endorsement of the recreation area by all three agencies was expected but necessary. The Department of the Interior report was the most interesting since it alluded to an estimated annual visitation of more than ten million. The number of estimated visitors had doubled since the *Delaware River Basin Report* had first been published.

Hearings on H.R. 89 and related bills were held on March 1, 1965, by the Subcommittee on National Parks and Recreation. As before, overwhelming support for the recreation area was heard. A resident of the Minisink Valley, Nancy Shukaitis, testified at the hearing to express opposition to the federal park. She called for a locally based conservation and land-use program to preserve the scenery of the area, with the federal government limited to an oversight role. She also requested that the subcommittee hold hearings in the Tocks Island region to hear from the people most affected by the proposal. Although the subcommittee was reluctant to do this, an additional hearing was held on April 22 in East Stroudsburg.

On May 20, 1965, the Tocks Island National Recreation Area bill was reported favorably out of the House Committee on Interior and Insular Affairs. Among the minor changes made to the bill was a name change. At the April 22 hearing, the Delaware Water Gap Chamber of Commerce had recommended that the name be changed to the Delaware Water Gap National Recreation Area. Certainly, the water gap was more famous than a small island that would soon disappear underneath a reservoir. The bill's writers, however, had garbled the name change, calling it the Delaware Valley National Recreation Area. The House passed the bill with its incorrect name on July 12.

The Senate Interior Committee reported out the House bill on August 13. The Senate committee corrected the name-change mistake and recommended that the Delaware Water Gap National Recreation Area be established. Also added was a proviso that the area was not to be extended into New York State unless specifically authorized by Congress. This provision was directed at persons lobbying to have the recreation area extended into the Neversink Valley.

The Senate committee's report was acted upon immediately. On the same day that the bill was reported out, the Senate passed it. Four days later, the House concurred with the Senate's changes. President Johnson signed the bill into law on September 1, 1965, making the Delaware Water Gap National Recreation Area a reality.

Public Law 89-158, which created the Delaware Water Gap National Recreation Area, was quite simple—only two and a half pages long. The law gave the Secretary of the Interior $37.4 million to acquire approximately 47,675 acres of land and $18.2 million to construct recreation facilities. The 47,675 acres represented additional federal land acquisition above the 24,000 acres authorized previously for the Tocks Island Dam Project. Most of the 72,000 acres that would make up the Delaware Water Gap National Recreation Area (DWGNRA) was privately owned. This land would be acquired by purchase, or, failing that, by condemnation. The federal government had never before attempted to create a national park by taking so much private land from its citizens. The job of acquiring the land was given to the Corps of Engineers since it would be acquiring land for the dam and reservoir project anyway.

Soon after the Delaware Water Gap National Recreation Area was established, the National Park Service opened a local office in East Stroudsburg. Peter DeGelleke, who in 1957 had first proposed the federal recreation area, was appointed planner-in-charge. A sketch plan of the recreation area had been presented in the *Delaware River Basin Report*. Ten major development areas, or parks within a park, were envisioned. The first job of the park service would be to flesh out the details of the sketch plan into a master plan for the recreation area's development.

The *Master Plan for the Delaware Water Gap National Recreation Area* was released in late 1966. The plan divided the original ten development areas into smaller units, or development sites. Development sites were areas containing a beach, a picnic area, a scenic overlook, or a boat-launch facility. The thirty-one largest sites focused on reservoir recreation. At these intensely used sites, the planners envisioned something on the order of 11,000 picnic tables, 6,500 camping sites, 135 boat ramps, 1,860 boat docks, 33,000 parking spaces, 15 food-service areas, and beaches for 66,000 bathers. In between all the recreation sites were acres of woodlands, open fields, and other scenic attractions.

The unifying theme of the master plan was the "parks within a park" concept. The various recreation sites were to be relatively self-contained in terms of recreational opportunities. With nine major beach areas, thirty-one camping areas, boat rentals, miles of hiking and bicycle trails, horseback riding, picnicking, hunting, fishing, sailing, motorboating, canoeing, rock climbing, winter sports, nature centers, playgrounds, ballfields, historical sites, interpretive facilities (including Tocks Island Dam itself), and all the rest, the DWGNRA promised to have something for everyone.

The Delaware Water Gap National Recreation Area was being designed to handle up to 150,000 visitors per day, or 10.5 million per year. Half of the areas administered by the National Park Service did not see 150,000 visitors in one year! The national recreation area around Tocks Island Reservoir would be the busiest park in the National Park System, the largest east of the Rocky Mountains, the first east of the Mississippi River, and the first to be developed around a Corps of Engineers' dam project. Acquiring the land for both the reservoir and recreation-area projects would be one of the largest real-estate activities ever undertaken by the Corps.

Land acquisition for the recreation area began in 1967, and by the following year the National Park Service had a visitors center and scenic overlooks in the Delaware Water Gap. At the end of its first year, the DWGNRA had recorded 24,000 visitors. Meanwhile, planning and land acquisition for the recreation area were continuing.

The Delaware Water Gap National Recreation Area was also beset by money problems as soon as it was authorized. The recreation area was affected by cutbacks due to the Vietnam War, but to a lesser extent than the dam project. The price of land was the major dollar problem for the National Park

Service. The agency had been authorized only $37,410,000 for the purchase of 47,675 acres of land. This would have been fine, except that the price of land in the Tocks Island region was far from stable. The massive amount of federal activity in the region, plus normal growth pressures, was driving the price of land ever upward.

The rising land costs (or speculation, depending on interpretation) were connected with several large, active real-estate developments in the Tocks region and the threat of more to come. Land developers in the DWGNRA were characterized by the pro-Tocks forces as speculators, hucksters, and fast-buck artists, and some of them were. The Hidden Lakes development, for example, had been started shortly after the DWGNRA had first been proposed. Critics of the DWGNRA were quick to point out that the president of the WRA/DRB was associated with this development even as his organization was lobbying for the park. Other developments wantonly chewed up valuable hemlock groves and other unique natural features, even after the recreation area was assured.

Most land developers, however, had been active before the DWGNRA was proposed. A considerable amount of money had been invested in land, roads, lakes, and other facilities. It was not logical that any of the developers would voluntarily suspend land sales just because of a park idea that was slow in coming. In fact, when the developer of Blue Mountain Lakes approached the federal government asking what he should do, the answer was to continue business as usual.[12] Government representatives were unable to predict when lands would be purchased and could not issue letters of intent that developers could use as collateral for business loans elsewhere. Local dam opponents considered the concern for land speculation to be a red herring designed to speed up congressional appropriations.

By 1969 it was clear that the original estimates for the DWGNRA were too low. The total cost of the park was now estimated to be at least $65.8 million, with land acquisition estimated to be at least $56.1 million. In an April 1970 report, the General Accounting Office released a nationwide study of the cost problems of park land acquisition. The report was highly critical of the DWGNRA because 29 percent of the National Park Service's land-acquisition funds had been spent acquiring 9 percent of the land proposed for the park. The GAO felt that the National Park Service could have saved a good deal of money by adjusting the park boundaries to exclude the active real-estate developments. Most were on the fringes of the park, anyway.

By 1975 the price tag for the DWGNRA was well over $100 million.[13] Congress dutifully raised the $37,410,000 ceiling to $65 million in 1972 and finally removed it altogether in 1978. Today, private in-holdings in the DWGNRA remain, but land acquisition is not being pursued aggressively.

Funding and land acquisition were internal problems. Externally, the establishment of the Delaware Water Gap National Recreation Area created concern for the secondary impacts on the surrounding region. The 10.5 million
visitors predicted by the National Park Service would need a wide variety of
public services and facilities. Their trips to the Tocks Island region would also
create a boom in gas stations, motels, restaurants, retail stores, tourist traps,
vacation homes, and other businesses, which, in turn, would bring new
permanent residents and other businesses to the area. The rural counties of
the Tocks Island region were not ready for this onslaught.

Shortly after Tocks Island Dam was authorized, planners began wrestling
with the secondary impacts. During the next ten years, probably close to a
million dollars was spent on planning studies by agencies of New Jersey, New
York, and Pennsylvania, and by the Delaware River Basin Commission, the
Tocks Island Regional Advisory Council (TIRAC), and others. TIRAC
(1965 to 1974), a seven-county agency, was organized by the WRA/DRB
specifically to address the impacts of Tocks Island Dam on the region. The
secondary impacts would become a major issue in the Tocks Island Dam
controversy.

When the many plans were completed, a truly amazing picture of the
secondary impacts of Tocks Island Dam and the recreation area emerged. It
looked something like this:

1. *Land Use Control.* According to everyone's projections, the Tocks Island
   region was going to grow even without the stimulus of federal activity in
   the region. Almost 120,000 new homes were projected in the four counties surrounding the reservoir by 1985. The economic impact from the
   park itself was estimated to be $28.5 million annually, a level of spending
   that could support 40 to 80 new restaurants, 2 to 3 grocery stores, 50 to 95
   motels, 25 to 50 gas stations, and 35 to 60 other establishments, ranging
   from sporting-goods stores to summer theaters.[14]

   The planners estimated that new development would consume 275
   square miles of vacant land by the year 2000.[15] The new development
   could be accommodated, but the lack of zoning controls in the area could
   lead to "another Coney Island with all the garish, crowded jumble that
   Coney Island connotes."[16] TIRAC, county, and state planning agencies
   tried but could never overcome the local apathy toward land-use planning.

2. *Highways.* With the exception of the new interstate highways, most highways in the Tocks Island region were designed for rural traffic loads. The
   thousands of people traveling to the park would clog these roads plus many
   of the connecting highways between the park and Philadelphia or the
   New York City/North Jersey area. Even the interstate highways appeared

vulnerable to traffic jams during peak vacation travel periods or weekend travel times.

Many new highways were recommended as a result of the recreation area. If all the highways had been built, residents of Philadelphia, Trenton, North Jersey, and New York City would have had several alternative multilane, high-speed routes to the Tocks Island region. In the Tocks region itself, they would have had new four- and six-lane highways to distribute them in all directions. Many of the new highways were not merely upgrades of existing highways. New Jersey's Foothills Highway, for example, would have cut across the farms and woods along the eastern base of Kittatinny Mountain. The upgraded and relocated U.S. Route 209 would have done the same thing in Pennsylvania. Access roads to the DWGNRA were planned off both of these new north-south highways.

In Pennsylvania the cost of expanding roads in the Tocks Island region was estimated in 1967 to be $50 million. This cost figure did not include proposed expressways from Philadelphia or the cost of upgrading U.S. Route 209, which was being relocated because of the dam project.[17] The New Jersey price tag for Tocks-related roads was much higher because it did reflect the costs of all the planned roads. In 1971 consultants to the state estimated a need for 183 miles of expressways costing over $685 million.[18]

3. *Waste Disposal.* With visitors and growth in the area came the need for new water supplies, solid-waste disposal, and sewage facilities. Water supply was not considered critical, even though an increase from 24 to 139 mgd was predicted.[19] Tocks Island Reservoir would be nearby if needed. More critical was solid waste. Solid-waste volumes in the region were predicted to rise from 53.5 tons per day to 3,890 tons per day, not counting the 2,500 tons per year from the park visitors.[20] Eleven square miles of sanitary landfills were needed to handle the loads from the hundreds of new garbage trucks.[21]

Sewage disposal was considered critical because it had the potential to cause water-quality problems in Tocks Island Reservoir. The amount of sewage in the region was expected to rise from 19.3 mgd to 99 mgd, with park visitors accounting for 5.7 mgd of the total.[22] The recommended plan for dealing with all the wastewater was to pipe and pump it to one or more large regional treatment plants. Sewage from up to 50 miles away would be piped to the treatment plant(s) via 565 miles of trunk lines, 275 pumping stations, and 5 trunk-line crossings of the Delaware River itself. The cost for such a massive system in the rural region was estimated to be $190 million.[23]

4. *Other Needs.* There were many other needs, almost too numerous to

mention. The recreation-wounded would need hospitals, doctors, nurses, paramedics, ambulances, and other health services. More police, more police cars, more jails, more courtrooms, more fire engines, more social workers, more traffic lights—more everything—would be needed. The Tocks Island region was being asked to substantially upgrade its public services to urban standards while expanding them tremendously at the same time.

The Delaware Water Gap National Recreation Area had seemed to be a truly positive addition to the Tocks Island Dam Project. Robert R. Nathan and Associates, the first consultant to examine the potential impacts of the project, referred to the park and reservoir as Central Park in Megalopolis. They believed that the DWGNRA would be analogous to Central Park in New York City. Central Park had originally been outside the developed area of New York City, but the city had soon developed around it. Nathan predicted that Megalopolis, the East Coast's super-city, would do the same with the Tocks Island region.

The many studies of the secondary impacts, however, highlighted the negative side of a lake-based recreation area. When faced with the potential impacts of the Tocks visitors, the planners proposed solutions and more solutions. To many persons, these solutions looked suspiciously like expensive problems. Secondary impacts would become a major issue in the Tocks Island Dam controversy.

9

Sunfish Pond: Prelude to Controversy

By itself, Tocks Island Dam was a large project. The Delaware Water Gap National Recreation Area made the dam project larger, even grandiose. Big as these activities were, there was still a third part to the Tocks Island Dam Project. The Kittatinny Mountain Project called for the construction of a pumped-storage system on top of Kittatinny Ridge by three power companies. Although a private development, the power project was intimately intertwined with the dam project. The power companies' proposal, however, destroyed a small mountaintop lake called Sunfish Pond. The efforts to save Sunfish Pond would lead directly to concerns about the environmental impacts of Tocks Island Dam itself.

Sunfish Pond is a forty-four-acre pond located on the top of Kittatinny Mountain in New Jersey, about one thousand feet above the Delaware River. Immediately beyond the western edge of the pond, Kittatinny Mountain begins its steep drop to the Delaware River. On a sunny day, Sunfish Pond and its one and a half miles of shore are spectacularly beautiful, with hardwood trees and scenic rock outcroppings reflected in the water. The Appalachian Trail skirts the western shore, and the pond is often the destination of day-hikers from the Delaware Water Gap. For hikers of the Appalachian Trail, Sunfish Pond is often remembered as one of the prettiest spots between Maine and Georgia.

Sunfish Pond and 708 acres of the surrounding land were purchased by Charles C. Worthington in 1890.[1] Worthington operated a resort in Shaw-

nee, Pennsylvania, and was a noted New Jersey industrialist. Resort visitors
made the pond the destination of their picnics, and after 1912 a water line
connected Sunfish Pond with the inn.[2] Through the years, however, Sunfish
Pond remained in its natural state as part of the Worthington family's 6,500-
acre game preserve on the New Jersey side of the Delaware River.

The Worthington Estate, or Tract, remained in the Worthington family
into the early 1950s, when the family decided to sell their landholdings. In
1952, therefore, the Worthington Tract caretaker, Walter van Campen,
contacted Robert B. Meyner about state purchase of the land.[3] Meyner was
the former state senator from the area and was about to make a bid for
governor. Van Campen gave Meyner a tour of the Sunfish Pond area on
January 1, 1953, and in the following November, Robert Meyner was elected
governor of New Jersey. After the election, he asked the Worthington family
to withhold their land from the market while he sought state money to
purchase it.

The effort was successful, and in 1954 the two sons of Charles Worthington sold the Worthington Tract to the State of New Jersey for $420,000. The
state established the area as Worthington State Forest, built campgrounds,
and otherwise left the land alone—with one exception. In September 1960
state fisheries personnel put poison in Sunfish Pond as a prelude to trout
stocking.[4] Approximately a thousand pounds of native fish were killed.[5] The
state stocked the pond twice with brown trout, but both attempts failed. As a
result, Sunfish Pond was virtually sterile and considered of little value by the
state's recreation (i.e., trout fishing) experts. Although some persons would
try later to depict Sunfish Pond as naturally sterile, this was definitely not
true.

The first examination of Kittatinny Ridge as a possible pumped-storage site
was made by the Public Service Electric and Gas Company in 1947.[6] Pumped
storage, however, was not technically feasible at that time. The idea showed
promise, however, because it solved one of the power industry's most critical
problems, the need to have enough generating capacity when power demand
was at its highest. Pumped storage solved this problem by using a power
company's idle generating capacity during off-peak periods to pump water
from a lower reservoir to a higher one. When high power demands required,
the water in the higher reservoir was returned to the lower one. In the
process, the fall of the water was used to generate electricity. Situated as it
was above the soon-to-be-built Tocks Island Reservoir, Kittatinny Ridge was
an ideal location for a pumped-storage facility.

The development of large reversible-pump turbines in the 1950s awakened
interest in the power potential of Kittatinny Ridge. In 1956 studies were
initiated by Public Service Electric and Gas, Jersey Central Power and Light

Company, and the New Jersey Power and Light Company, three of New Jersey's largest power companies. Undoubtedly adding impetus to the studies was the Corps' favorable report on the construction of Tocks Island Dam in the same year.[7]

The studies of Kittatinny Mountain determined that suitable upper-reservoir sites existed on the ridge. Two sites for the lower reservoirs were also found, one east of the ridge and the other to the west. The eastern lower reservoir could be created by damming Yards Creek, a small tributary of the Paulins Kill. The western lower reservoir was Tocks Island Reservoir. A three-phased project evolved out of the studies. The first phase involved the construction of the Yards Creek side of the project, which consisted of an upper reservoir east of Sunfish Pond and a lower reservoir on Yards Creek. Phases two and three of the Kittatinny Mountain Project found more reservoirs built on top of the mountain and conduits sent down the mountainside to the Delaware River and Tocks Island Reservoir.

The interest in pumped storage stimulated the formation in 1959 of a Power Work Group to advise the Corps' Delaware River Basin planning program.[8] The group consisted of the Corps of Engineers, the Federal Power Commission, and the six major Pennsylvania, New Jersey, and New York power companies. Various potential pumped-storage sites were examined by the committee, including facilities at dam sites other than Tocks Island. The construction of pumped-storage facilities, however, was kept out of the Corps' recommended plan and therefore was never authorized by Congress.

On April 13, 1960, the three power companies publicly announced the Kittatinny Mountain Project at a Corps of Engineers public hearing.[9] Not only were the companies prepared to handle *all* power development for Tocks Island Dam, but they were willing to allow their pumped-storage facilities to be used for diverting Tocks Island water to North Jersey. Under this proposal, water would be pumped over Kittatinny Mountain from Tocks Island Reservoir to Yards Creek, where it would then be piped to North Jersey. The use of the pumped-storage facilities was alleged to be cheaper than any other New Jersey alternative for getting Tocks water. Although New Jersey would remain undecided about the proposal, in 1963 the City of Newark purchased an abandoned railroad right-of-way running from near Yards Creek to Sparta for the water pipeline.[10]

The New Jersey water-supply proposal was a carrot offered to New Jersey. Construction of the Kittatinny Mountain Project required state-owned land on top of Kittatinny Ridge in Worthington State Forest. In 1958 the power companies asked state officials to sell them the needed land, and a sales agreement was negotiated by March 1960. Most people, however, remained ignorant of the pending sale. Six months after the sale was supposedly an-

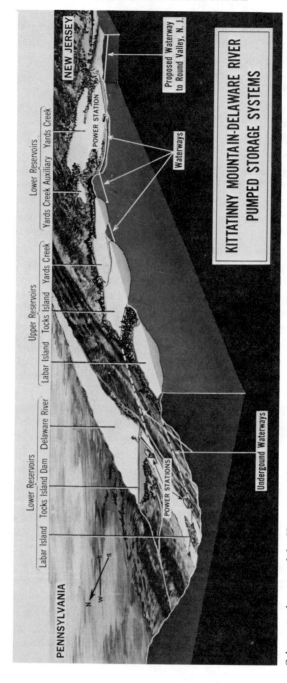

Schematic drawing of the Kittatinny Mountain Pumped-Storage Project. The Yards Creek portion of the project was built first. Note the proposed water transfer to North Jersey via Round Valley Reservoir, one of several schemes for getting Tocks water to North Jersey. (Jersey Central Power and Light Company)

nounced, New Jersey fisheries personnel began their trout-stocking efforts in Sunfish Pond. This suggests that even employees of the Department of Conservation and Economic Development were unaware of the pending sale by their department.

The sale of Sunfish Pond and surrounding land occurred on March 1, 1961. New Jersey received $250,000 and 125 acres of land, including Delaware River frontage that the state wanted for recreation. The power companies, in turn, received 785 acres of land, including Sunfish Pond, plus easement rights for roads, penstocks, and transmission lines. Who got the best of the deal would be argued again and again in the subsequent controversy.

The Yards Creek phase of the project was built first since it did not depend on the construction of Tocks Island Dam. On May 21, 1962, the project was sent to the Delaware River Basin Commission for review under Section 3.8 of the Delaware River Basin Compact.[11] A Federal Power Commission permit was obtained on March 14, 1963, and the Yards Creek Pumped-Storage Generating Station became operational in 1965.

Shortly after the Yards Creek Project received its federal permit, the power companies applied to the Delaware River Basin Commission for approval of the Tocks Island phases of the Kittatinny Mountain Pumped-Storage Project. The proposal called for two more reservoirs on top of Kittatinny Mountain. Drowned and bulldozed under one of these were to be Sunfish Pond and its bucolic setting.

Although the power companies requested a prompt decision on their application, the Delaware River Basin Commission did not give it. Pumped storage involving Tocks Island Dam contained policy implications that were absent in the Yards Creek proposal. First of all, DRBC had to decide whether it wanted to build the project itself. Early in its history, the commission began studying ways to get into the hydropower business. DRBC was interested in the possibility of building and operating a power project not only at Tocks Island but also at Beltzville Dam. It figured that the benefits of power development at Corps dams should accrue to the public, i.e., the commission.

By late 1965, the three New Jersey power companies had invested nine years of effort and a considerable amount of money in the Kittatinny Mountain Project. The Yards Creek portion of the project was done, but the Delaware River Basin Commission had yet to grant permission for the other two phases. By now, DRBC and the power companies were negotiating the amount of money the former would get for "renting" its water in Tocks Island Dam. The power companies would also be expected to pay part of the Tocks Island Dam Project costs as well.

Through 1965 the public remained generally ignorant of the Kittatinny Mountain Project and its impacts. In that year, however, two Warren

County, New Jersey, residents began independent attempts to save Sunfish Pond. Glenn Fisher was probably the first to call public attention to the pending destruction of Sunfish Pond. Fisher, a former U.S. Department of Agriculture employee, pointed out the pending loss at the April 22, 1965, congressional hearing in East Stroudsburg. Soon after, he began writing protest letters and collecting signatures along the Appalachian Trail in an effort to save the pond. Independent of Fisher, Casey Kays began his own effort to save Sunfish Pond around the same time. Kays was a Hackettstown resident with strong feelings about Sunfish Pond and its pending destruction. In 1965 he began a massive letter-writing campaign to local newspapers and anyone else who might be interested. Others joined in the fight, and a small network of persons opposed to the pumped-storage project was created.

In January 1966 Glenn Fisher organized the Lenni Lenape League and began sending a small newsletter, *Lenape Smoke Signals*, to other foes of the pumped-storage project. Fisher and Kays met for the first time on April 19, 1966, when they and four others got together to plot strategy. The group decided to test support for their Sunfish Pond efforts. On May 8 a pilgrimage to the pond was conducted by the Lenni Lenape League. In spite of rain, the short notice, and the fact that it was Mother's Day, the league collected 655 signatures for its petition. With this initial success, the fight to save Sunfish Pond began in earnest.

The massive letter-writing campaign and the publicity surrounding the pilgrimage soon reached all levels of government. The "Save Sunfish Pond" fight presented a dilemma for the pro-Tockers. The pumped-storage project magnified the benefits of Tocks Island Dam and, along with the recreation area, made the Tocks Island Project a super-project. While exciting from a water-resources viewpoint, pumped storage was, nevertheless, industrial development. It conflicted with the purpose of the national recreation area, destroyed Sunfish Pond, and affected the integrity of the Appalachian Trail. As a result, many conservation organizations in New Jersey began taking a hard look at what was going on in the Tocks Island region.

Help was soon to come from another source as well. The National Park Service had established a Scientific and Educational Advisory Committee to help it plan the recreation area. Among the members of the committee was Professor Francis J. Trembley, a highly respected ecology professor at Lehigh University. In Trembley's opinion, fluctuations in Tocks Island Reservoir caused by the pumped-storage operations would harm fish spawning in Tocks Island Reservoir. By the end of 1966, the important New Jersey State Federation of Sportsmen's Clubs, representing 450 clubs, was petitioning the Delaware River Basin Commission to study the spawning question. For those not interested in fishing, the reservoir fluctuations raised the specter of mudflats ringing Tocks Island Reservoir, or fears that the dikes surrounding the upper reservoirs would be visible from the valley below.

The National Park Service soon broke with the other pro-Tocks agencies over the pumped-storage issue. In its 1966 *Master Plan,* the National Park Service stated, "Sunfish Pond is proposed for use as a pump storage reservoir by Public Utilities, a use which would deface the landscape and create a barrier across Kittatinny Ridge between Catfish Pond and the Water Gap."[12] Soon various local-government entities joined the park service in opposing the pumped-storage project in the DWGNRA. The *Bethlehem Globe-Times,* in a November 21, 1966, editorial, summed up the controversy by stating that if pumped storage ruined the recreational values of Tocks Island the result would be "one of the biggest hoaxes ever perpetrated in the name of the common good."

The campaign of letters and words continued unabated in 1967. In early 1967, however, the Delaware River Basin Commission began a three-year research study of the effects of pumped storage on fish. The study was developed by a committee of state and federal fisheries biologists. The three power companies were required to pay $50,000 per year and to provide research facilities. Four two-acre ponds were built at the Yards Creek power plant and stocked with various species of fish. During the day the ponds were subjected to rises and falls similar to what might be expected from pumped-storage operations.

The Lenni Lenape League, meanwhile, was growing, reaching about five hundred members in 1967. Glenn Fisher was the executive director, and Casey Kays was the president. From their pens and others, literally hundreds of letters were sent to governors, state and federal officials, legislators, U.S. senators, newspapers, and others throughout New Jersey and eastern Pennsylvania. Soon opposition voices were being heard in the New Jersey legislature calling for public inquiries concerning the original sale of the pond. In Congress the long-time supporters of Tocks Island Dam became concerned. At least one, Representative Frank Thompson, expressed outright opposition to the pumped-storage idea.[13]

On May 7, 1967, the Lenni Lenape League sponsored its second pilgrimage to Sunfish Pond. Although thousands of supporters were predicted, the trek was hampered by heavy rains, winds, and cold temperatures. In spite of this, more than two hundred people turned out for the hike. The league was not discouraged by this turn of events. They had received a letter from U.S. Supreme Court Justice William O. Douglas expressing interest in their cause and a hike up to Sunfish Pond. A second 1967 pilgrimage was quickly arranged.

The Douglas hike to Sunfish Pond was the high point of the Lenni Lenape

League's publicity efforts. It occurred on June 17 under much better weather conditions than the earlier 1967 hike. Because of the publicity, crowds estimated to range from four hundred to one thousand turned out. William O. Douglas at this time was sixty-eight, and his age and young wife made the hike a true media event. Douglas, moreover, was identified nationally with environmental causes and had hiked the entire Appalachian Trail in 1959. Photos of the justice in his battered hat, old shirt, and baggy pants made good copy. Quotes from Douglas found their way into the Sunday papers read by millions of people the following day. His opinion was quite clear: "I wish to identify myself strongly with this cause. Sunfish Pond is a unique spot and deserves to be preserved."[14]

Also drawing widespread news coverage in 1967 was the August 17 public hearing of the Delaware River Basin Commission. By this time DRBC and the power companies had reached a tentative agreement allowing the latter to develop the Tocks Island pumped-storage facilities. The Delaware River Basin Commission was willing to "rent" its Tocks Island water for $500,000 per year. In addition, the power companies would have to pay a share of the costs of the dam and give the Delaware River Basin Commission an equivalent of 281.5 million kilowatt-hours of electricity each year. With this power, the commission envisioned pumping sewage and water up and down various watersheds for water-pollution control and water-supply purposes.

The public hearing was marked by emotional arguments on both sides of the controversy. Lengthy testimony by the power companies made a persuasive case for the benefits that the public would receive from the project. Joining the opponents was the Sierra Club, the first national environmental organization to enter the fight for Sunfish Pond. Most attention, however, was focused on the clash between former Warren County state senator Warren Dumont, Jr., and former governor Robert B. Meyner. Dumont alleged that the sale of Sunfish Pond had been unusually secretive and that state officials had been too cooperative with the power companies. Meyner took exception to these allegations. Both politicians had been long-time political foes in Warren County, and the sparks at the hearing made Sunfish Pond headlines in newspapers all over New Jersey and eastern Pennsylvania.

At the August 17, 1967, hearing and at other times, the power companies stated that there were no alternatives to the destruction of Sunfish Pond. After the hearing, however, DRBC's staff advised the power companies that they should develop an alternative to the use of Sunfish Pond.[15] It was becoming increasingly important that the mess get cleaned up before it cast doubts on the Tocks Island Dam Project itself.

By mid-1968 the fight to save Sunfish Pond was reaching its zenith. The Lenni Lenape League's fourth annual pilgrimage to the pond drew at least two

Supreme Court Justice William O. Douglas at the shores of Sunfish Pond on June 17, 1967. (Albert Dillahunty, National Park Service Collection)

thousand hikers. In the New Jersey legislature, movements were under way to authorize an investigation of the sale of the pond to the utility companies and to buy back the pond. The latter proposal was largely identified with Assemblyman Thomas H. Kean, a future governor of New Jersey. The adverse publicity greatly concerned New Jersey's congressional delegation since they would be expected to sponsor any bill authorizing private power development at Tocks Island Dam.

In July 1968 a compromise was reached between the Delaware River Basin Commission and the three power companies. The power companies indicated that they would give Sunfish Pond back to the state in trade for one hundred acres of nearby land. Use of Sunfish Pond would be avoided by raising the height of the Tocks Island Upper Reservoir by seventeen feet. The battle for Sunfish Pond appeared to be over—as far as the supporters of Tocks Island Dam were concerned.

On October 22, 1968, the Delaware River Basin Commission amended its comprehensive plan to allow the development of pumped storage as part of the Tocks Island Dam Project.[16] The special conditions of the commission's action called for the preservation of Sunfish Pond, the landscaping of the upper reservoir dikes, and the underground placement of all penstocks, generating plants, transmission lines, and other facilities. Operation of the project was to be integrated with the operation of Tocks Island Reservoir, with the possibility of future mitigating measures being required when DRBC's fisheries research project at Yards Creek was completed.

After the Delaware River Basin Commission decision, action on the pumped-storage proposal shifted to Congress. By the end of 1969, a bill to deauthorize the original, conventional power development at Tocks Island and add private pumped-storage power had passed the Senate. By June 1970 pumped-storage power was an authorized purpose of the Tocks Island Dam Project.

The supporters of Tocks Island Dam had assumed that the compromise saving Sunfish Pond would end the controversy. They were wrong. A new rallying cry, "Save All of Sunfish Pond Including Its Watershed," reflected a general concern that the enlarged Tocks Island Upper Reservoir would leak into Sunfish Pond and destroy its naturalness. Part of this concern stemmed from the fact that the existing Yards Creek Upper Reservoir leaked. Publications from the Lenni Lenape League pictured an algae-filled Sunfish Pond with two hundred-foot-tall dikes towering above it, the Appalachian Trail pushed off the ridge, and other horrors.

By this time the Sunfish Pond controversy had evolved into something larger. As early as August 1968, Glenn Fisher and others were writing letters talking not only about the potential harm to the pond but also about the loss of the free-flowing Delaware River, the loss of river bottomlands, the impact on ecology, water quality, and other issues.[17] These were impacts anticipated from the construction of Tocks Island Dam and not from the pumped-storage facilities.

In April 1969 the Lenni Lenape League received the 1969 *Holiday Magazine* Award for a Beautiful America for its Sunfish Pond fight. In the following month, the league and the North Jersey Group of the Sierra Club sponsored another pilgrimage to the pond. More than nineteen hundred persons attended the hike, even though eight months of widespread publicity had credited the Delaware River Basin Commission with saving the pond. By 1970 the pilgrimages were attracting people who were less interested in saving Sunfish Pond's watershed than they were in stopping Tocks Island Dam.[18] The 1971 pilgrimage found thousands jamming the narrow Old Mine Road and other trail access points. At the hike, the Lenni Lenape League, the Delaware Valley Conservation Association, the Sierra Club, and others were out in force to solicit support for New Jersey Assembly Bill A-517. This bill

Collecting signatures at the fall 1969 hike to "Save Sunfish Pond." (Albert Dillahunty, National Park Service Collection)

called for the federal government to "restudy the wisdom of Tocks Island" and to delay the pumped-storage project until the study was done.

Sunfish Pond, meanwhile, was again owned by the State of New Jersey. On July 1, 1969, New Jersey Governor Richard J. Hughes accepted a deed for the forty-four-acre pond and sixty-eight acres of surrounding woods from Edwin H. Snyder, chairman of Public Service Electric and Gas, and Ralph F. Boliver, president of Jersey Central Power and Light. The pond again became part of Worthington State Forest. Former state lands to the east of Sunfish Pond remained in the hands of the power companies for the eventual construction of the upper reservoir. Sunfish Pond was no longer needed.

In the spring of 1970, the Interior Department announced that Sunfish Pond would be included in the National Registry of Natural Landmarks. On March 19, 1973, a bronze plaque mounted on a stone marker was dedicated at the pond. Among the speakers at the ceremony were Casey Kays and Assemblyman Thomas Kean. Both had contributed to the success of the "Save Sunfish Pond" fight. Later, the trail to Sunfish Pond from the Worthington State Forest headquarters was named the Douglas Trail in honor of Justice

Douglas and his famous hike. Today Sunfish Pond is still a lovely spot on top of the mountain.

The Save Sunfish Pond fight made thousands of people aware of the large construction projects threatening the natural beauty of the Tocks Island region. Ultimately, the concern for a small pond evolved into a broader concern for the environmental impacts of Tocks Island Dam. As a result, the Sunfish Pond controversy became a warm-up for a much bigger fight to stop Tocks Island Dam.

10

Tocks Island Logjam

Because of the delay in the start of construction, Tocks Island Dam got caught in a cross fire. Public interest in preserving the environment had been increasing ever since the dam was authorized. Nationally, this interest had culminated in the passage of some key environmental legislation, such as the National Environmental Policy Act. By the 1970s, Tocks Island Dam would have been scrutinized closely by the environmentalists, if for no other reason than that its sponsor was the hated U.S. Army Corps of Engineers. The National Environmental Policy Act, however, would create even larger headaches for the Corps.

The Corps of Engineers, like most federal agencies, responded quickly to the environmental impact statement requirements of the National Environmental Policy Act (NEPA). The preparation of environmental impact statements, however, represented uncharted waters in 1970. The Council on Environmental Quality (CEQ) had published some guidelines for preparing statements, but nobody knew how comprehensive the statements were to be. It was inconceivable that a project like Tocks Island Dam could be seriously affected by the new requirements. It had been nursed along for seven years, enjoyed the support of the four Delaware River Basin states, and was about to begin construction. More than $25 million had already been spent in planning, design, and land acquisition.

The Corps' first attempt at preparing an environmental impact statement for Tocks Island Dam was a very preliminary seven-page statement. The statement was distributed for comments in October 1970 and submitted to the CEQ in February 1971, even as the Corps was working on a more comprehensive report. The Corps never expected the short statement to meet the NEPA requirements. Its size, its apparent downplaying of environmental concerns, and its cursory treatment of environmental impacts, however, heightened suspicions about the Corps' sincerity.

CEQ's comments on the Tocks Island Dam environmental impact statement were extensive.[1] Overall, CEQ felt that the statement lacked depth in its environmental analysis. They suggested that an independent organization, such as the National Academy of Sciences, prepare the statement for the Corps. A coordinated, multidisciplinary, multiagency report covering all aspects of Tocks Island Dam was also recommended. Such a report would cover not only Tocks Island Dam but also the Delaware Water Gap National Recreation Area and the Kittatinny Mountain Pumped-Storage Project.

CEQ also presented specific comments concerning subjects that they felt should be addressed in detail. CEQ wanted discussions of:

1. Water quality in the reservoir, particularly the potential for the reservoir to become eutrophic;
2. Alternatives to the dam (the Corps' alternatives section had tended to justify the dam project rather than to consider that alternatives might exist);
3. Fisheries, particularly the passage of shad and the effect of the reservoir operations on fish habitats both in the reservoir and downstream;
4. Economic and social trade-offs, including secondary costs and benefits, land-use control, and the trade-offs associated with a free-flowing river versus an impounded one;
5. The impacts of seasonal fluctuations in reservoir levels on fisheries, recreation, and aesthetics;
6. The loss of wildlife habitat due to reservoir flooding; and,
7. The effect of siltation on fish habitat and the useful life of the reservoir.

Of the many concerns raised by the CEQ, the potential eutrophication of Tocks Island Reservoir was the most serious. CEQ believed that "eutrophication of the proposed Tocks Island Reservoir because of high nutrient runoff from the Delaware River watershed is a real possibility a few years after the dam's completion. At best this eutrophication will only destroy the reservoir's proposed game fishery resources; at worst it will make the reservoir unavailable for recreation."[2] These were pretty damaging comments, considering that recreational benefits made the Tocks Island Dam Project economically feasible.

Adding to the potentially dismal recreation picture were CEQ's concerns about the impact of fluctuating reservoir levels and the potential for wide-spread mudflats due to reservoir drawdowns. Although the Corps attempted to minimize the mudflat problem, even they admitted that the operation of Tocks Island Reservoir could vary the pool level from two to fifty-nine feet (the latter during a one-hundred-year drought). During a normal rainfall year, drawdown was estimated to be seven feet during the recreation season and eighteen feet for the year.[3]

The Corps' immediate response to the CEQ comments was to postpone the start of dam construction,[4] a request of the council. To the pro-Tockers, the postponement was considered temporary since the Corps was already hard at work on a more comprehensive environmental statement.

The environmental questions about Tocks Island reached the floor of Congress when one of the Delaware River Basin congressmen began actively opposing the dam. In July 1971 Representative Pierre S. duPont IV of Delaware tried to strip the $3.7 million allocated by the Public Works Committee for the start of dam construction. DuPont's opposition stemmed from the environmental impact statement's cursory treatment of the impact of Tocks flows on downstream water quality, fisheries, and oyster production. These were Delaware issues. The freshman congressman also felt that Congress should not grant funds for a project that had unanswered environmental questions. In duPont's mind, "CEQ can only recommend sound environmental policy. It's up to Congress to make it a reality."[5]

The reaction to duPont's amendment was swift and vociferous. Frank Thompson of New Jersey and Fred Rooney of Pennsylvania rose immediately to criticize the amendment and Congressman duPont. Both were long-time proponents of Tocks Island Dam. Thompson stated that duPont's "zeal exceeds his information,"[6] and Thompson made even more critical remarks that were subsequently stricken from the record.[7] Adding to the criticism was James Wright, congressman from Texas and an influential public-works advocate. Wright reminded the House that the Book of Genesis had called upon man to subdue the Earth. He equated further study of Tocks Island Dam with Kipling's poetic description of old men: "They peck out, dissect, and extrude to the mind; The flaccid tissues of long-dead issues." Wright called upon his House colleagues "to vote down this amendment by a sound and emphatic vote."[8] He got his wish.

On October 1, 1971, the Corps filed its final environmental impact statement with the Council on Environmental Quality. The report was now more than one-hundred pages long and included supporting documents. As would be expected with a project the size of Tocks Island Dam, many adverse and permanent environmental impacts were identified. According to the Corps, these included:

1. *Impacts due to the creation of a thirty-seven-mile lake:* loss of a free-flowing river; loss of agricultural lands; reductions in fish spawning grounds; exposure of rock cuts; partial inundation of the historic Pahaquarry copper mines; loss of most of the historic Old Mine Road in the Minisink; reductions in wildlife and natural vegetation; loss of historically important buildings and sites; loss of archaeological sites; the need for floodwalls in the Port Jervis area; the acceleration of natural eutrophication; and hardships on current residents in the Minisink;

2. *Impacts due to reservoir operations:* fishery problems, including reductions in shad migration; the exposure of land areas due to reservoir drawdowns; and other potential problems associated with the latter; and,

3. *Impacts due to recreation:* traffic jams; solid waste disposal; and sewage collection and treatment.

Accompanying the environmental impact statement were two documents. The first, entitled *Tocks Island Lake Development: A Comprehensive Evaluation of Environmental Quality,* was the multiagency report that had been suggested by CEQ in its February comments. It attempted to tie together the environmental impacts of all three projects operating in the Tocks Island region: the dam, the recreation area, and the pumped-storage project.

The second and more important supporting document was *An Appraisal of the Potential for Cultural Eutrophication of Tocks Island Lake,* prepared for the Corps by Jack McCormick and Associates of Devon, Pennsylvania. The idea for a special study of the eutrophication issue had also been suggested by CEQ. McCormick, a terrestrial ecologist, assembled a temporary team of specialists, and in August and September 1971 they conducted a quick but detailed study of the eutrophication potential of Tocks Island Reservoir.

Eutrophication was to become the emotional issue of the Tocks Island Dam controversy. The term refers to the degree of aquatic plant activity in a water body. Eutrophication is a natural process that "ages" lakes and eventually converts them to swamps and meadows. The speed of the aging process is usually very slow in natural water bodies, but when humans enrich the environment with nutrients such as those found in sewage, fertilizers, livestock wastes, and detergents, the process accelerates. Symptoms of eutrophication include algae blooms and excessive rooted aquatic plants. These, in turn, lead to the disappearance of game fish, changes in water quality, and various aesthetic problems.

The potential eutrophication of a reservoir in the Tocks Island area had been first raised in the 1930s by the Lance brothers in their water-supply proposal for Philadelphia. In the mid-1960s, it was raised again by Dr. Francis J. Trembley, the ecology professor from Lehigh University. As early as February 1965, Trembley was sounding the alarm about the potential water-quality problems associated with ten million visitors to Tocks Island Reservoir. Later, he warned that Tocks Island Reservoir had the potential for becoming "one murky body of water,"[9] or "one gigantic cesspool."[10]

Early in its planning for Tocks Island Dam, the Delaware River Basin Commission had also recognized the potential for water-quality problems in the reservoir if sewage was not properly handled. In 1966, therefore, DRBC hired Roy F. Weston Engineers to develop a sewage-treatment plan for the area, the so-called Tocks Island Regional Environmental Study, or TIRES.

One of Weston's project engineers on the TIRES contract was Thomas H. Cahill, who actually developed many of the TIRES alternatives. Concurrent with his employment, Cahill was pursuing a master's degree in civil engineering at Villanova University. The subject of eutrophication interested him, and he developed a master's thesis project that combined his interest in eutrophication with his work on TIRES.

The objective of Cahill's thesis was the development of a mathematical model that could predict the potential of Tocks Island Reservoir to undergo accelerated eutrophication. Cahill hoped to use his model to evaluate the benefits of different sewage-treatment levels (i.e., the TIRES plan) for arresting eutrophication. Cahill's model, however, predicted that Tocks Island Reservoir would become eutrophic under *existing* conditions. The phosphorus levels in the free-flowing Delaware River were already "in excess of critical levels" due to agricultural runoff and other sources of nutrients.[11] Population growth and park visitors could increase nutrient levels and accelerate the eutrophication process even more. Cahill concluded that, "without positive action now, the chances are great that within a few short years, the clear lake will become clogged and polluted with masses of floating algae."[12]

Cahill's employer had encouraged his pursuit of a graduate degree. When his thesis was completed, however, the firm was uncertain whether to incorporate the thesis findings in the TIRES report.[13] Their client, the Delaware River Basin Commission, was highly sensitive to negative findings about Tocks Island Dam. After much internal debate, the firm included portions of Cahill's thesis as an appendix. It was in the form of an appendix to a technical report that the thesis was widely distributed.

The 1971 McCormick study was, in some respects, the second phase of Cahill's thesis. In fact, Tom Cahill had left Weston and was one of the specialists hired by McCormick for the study. The work was hampered by the same limited nutrient data that had originally hampered Cahill's thesis work, but some new data had been collected as a result of CEQ's comments on the environmental statement. These included a Delaware River Basin Commission study of the poultry industry in the drainage area above Tocks Island

Reservoir.[14] This study had found that the chicken population in the upper Delaware region, particularly in New York State, contributed phosphorus loadings equivalent to a human population that was four times the number of people living there. In addition to these data, the McCormick researchers had available a great deal of scientific research that had been completed nationwide after Cahill had done his thesis.

The McCormick study concluded that, "without positive and immediate action, the potential for the onset of accelerated cultural eutrophication during the early stages of reservoir operation is a real and serious threat to the full use of the impoundment for the multiple purposes cited for its justification."[15] The investigators recommended substantial reductions in all existing sources of nutrients, at least 98 percent phosphorus removal at Tocks Island regional sewage-treatment plants, and more studies.

The McCormick study was not the last eutrophication study of Tocks Island Reservoir. During the ensuing four years, at least ten other studies were conducted on aspects of Tocks eutrophication. Most concluded that eutrophication would occur to some degree unless phosphorus levels were reduced. How serious the problem would be, and whether it could be controlled, were questions that were never satisfactorily answered. One thing was sure, however. All sources of wastes in the Tocks drainage area would have to be cleaned up. Significantly, this included the cow and chicken manure wastes from New York State farms.

Although the McCormick study was not the last to address eutrophication, it did heighten CEQ's fears that Tocks Island Reservoir could have massive algae problems. The council, therefore, told the Corps of Engineers that more information was needed and that a specific plan for reducing nutrient loadings to the reservoir was needed. CEQ also wanted a substantial commitment from the Corps, the DRBC, and the states in which a nutrient-control plan in the Tocks Island drainage area would be implemented.

At this time, CEQ also continued to be very much concerned about the possible harmful effects of the reservoir fluctuations. This issue would eventually be overshadowed by the eutrophication issue, however. The Corps' position on reservoir fluctuations was understandable. They contended that reservoir fluctuations were inherent in any multipurpose dam project but that the impacts would be minor. The Corps also believed that most problems could be mitigated by modifying reservoir operations. During droughts, the need for water would greatly outweigh any harm caused by large reservoir drawdowns.

As the Corps was transmitting its environmental impact statement to the Council on Environmental Quality, it was also getting ready to advertise for construction. Tocks Island Dam had been ready for construction since the spring of 1971, and during the summer Congress had appropriated the necessary funds. The Corps had delayed the start of construction under the assumption that its October environmental impact statement would satisfy CEQ. CEQ's comments on the report, however, specifically requested that the Corps "not advertise for bids, or in any way initiate construction of the Tocks Island Dam until an Executive Office decision has been made relative to eutrophication and the drawdown problems."[16] Construction of Tocks Island Dam was halted before it began.

The Corps' problems in Congress increased as the fiscal year 1973 appropriations came around. In February 1972 the Chairman of the Council on Environmental Quality, Russell E. Train, wrote to Governor Rockefeller asking for assurances that New York State would clean up the wastes from its chicken and dairy farms. The nutrients caused no problems in the undammed Delaware River, but their control was believed essential for preventing eutrophication in a downstream reservoir. Train also sought assurances from the other three Delaware River Basin states, but he wanted special assurances from New York, since the cleanup of the manure problem would be its responsibility.

Governor Rockefeller's reply was released in time to affect the deliberations on Tocks Island funding.[17] New York, argued Rockefeller, would receive few benefits from the Tocks Island Reservoir. It was unfair to ask New York State to pay for nutrient controls when their costs should be charged to the dam project. Rockefeller promised to evaluate the needs of the upper Delaware with other pollution-control needs in the state. It did not take any imagination to realize that New York's massive urban-pollution problems would be the state's highest priority.

The lack of assurances from Rockefeller (as well as the other governors) spelled trouble for Tocks Island Dam's appropriations. In June the House Appropriations Committee approved $14.8 million for construction but stipulated that the funds could be used only for land acquisition until the assurances were given. The Senate Appropriations Committee concurred with this decision, and by September 1972 Congress had officially stopped the construction of Tocks Island Dam.

Support for Tocks Island Dam was eroding elsewhere, too. The Delaware River Basin Commission had always been a major supporter of the dam. Its staff fully backed the dam, and the representatives of the four state governors on the commission shared this feeling. As recently as 1970, DRBC had unanimously passed a resolution calling not only for construction of the dam but a speedup in construction to offset the Vietnam-related delays.[18]

In May 1972 the commission was holding a "summit" meeting, in which the four governors and the Secretary of the Interior met officially as the Delaware River Basin Commission. Tocks Island Dam was on the agenda.

The commission was concerned about rising public opposition to the project and also wanted to discuss the nutrient-control assurances sought by CEQ.

Representing New Jersey at the summit meeting was Governor William T. Cahill. Cahill had supported Tocks Island Dam as a congressman in the 1960s, but now he was not so sure. He told his fellow commissioners about his new reservations at the meeting. Cahill did not question his state's need for flood control, water supply, or the other benefits of the dam. What worried him were the dam's negative impacts: eutrophication, costly highways, costly sewage systems, and the like. He was unwilling to give any assurances to the CEQ or to support dam construction until New Jersey reevaluated its long-standing support for Tocks Island Dam. Cahill promised to report on his state's evaluation at the fall summit meeting.

Cahill's decision to restudy New Jersey's support of the dam project shocked the pro-Tockers. Between May and September, the supporters of Tocks Island Dam waited with some trepidation for the results. Throughout the period, the governor was constantly lobbied to support dam construction. Dam opponents, however, did little to counter this lobbying effort. Being under intense political pressure from the pro-Tocks faction, Cahill was disappointed by this lack of support.[19]

The actual work on New Jersey's reevaluation was performed by the state's Department of Environmental Protection under Richard J. Sullivan, a veteran of New Jersey's pollution-control programs. Aiding Sullivan was Thomas O'Neill, a graduate of Princeton University's Woodrow Wilson School. During the spring and summer, Sullivan and his staff labored to make some sense of Tocks Island Dam. Much of the available background material was biased or presented broad regional points of view rather than focusing on New Jersey issues. The governor wanted the reevaluation to consider only New Jersey's benefits, costs, and impacts, plus the ramifications of withdrawing the state's support for the dam project. It was not an easy task.

As Sullivan's staff was completing its analysis, Governor Cahill decided to visit the Tocks Island region. On September 9, 1972, he and Sullivan flew by helicopter over the Tocks Island Dam site and reservoir area. Down below them were acres of farmland, homes, and whole communities that were in the process of being converted into either a national recreation area or a reservoir bottom. It was sobering. Afterward Cahill met with local officials in Newton. At the meeting Cahill heard differing opinions about the merits of Tocks Island Dam and the recreation area. There was universal agreement on one thing, however. Local officials were afraid of being stuck with the tremendous costs associated with visitors to the national recreation area. They wanted federal funds to pay for the needed public services and facilities. Cahill left the meeting extremely concerned. The local officials had verified his staff's findings.

On September 13, 1972, Governor Cahill reported back to the Delaware River Basin Commission. Cahill told the other commissioners that New Jersey was not opposing Tocks Island Dam. New Jersey recognized that Tocks Island Dam provided significant flood-control and water-supply benefits to the state. New Jersey was still counting on Tocks for 300 mgd of water, and Cahill was very aware that Tropical Storm Agnes had wreaked havoc on the Susquehanna River Basin two months earlier.[20] The governor, however, had conditions that had to be satisfied before New Jersey would support dam construction. There were seven in all:

1. Legislation would have to be enacted in both New Jersey and Pennsylvania giving each the authority to regulate land use on floodplains;
2. New Jersey was to be left out of the large regional sewage-treatment plan adopted by DRBC for the Tocks Island region;
3. A plan would have to be developed to control nutrients from the area upstream of Tocks Island Reservoir in order to prevent eutrophication;
4. Legislation would have to be enacted giving New Jersey the authority to regulate land use in its portion of the Tocks Island region;
5. Substantial (preferably 90 percent) federal funding for new highway projects would have to be provided for access to the park;
6. A reduction would be necessary in the planned park-visitor load (from 10.5 million annual visitors to 4 million); and,
7. Federal payments would have to be made to the local governments to offset the local tax losses caused by the federal land-acquisition process.

The reactions to Cahill's conditions were mixed. New York and Delaware commissioners praised his position as statesmanlike and farsighted. There were many others, not connected to the DRBC, who criticized the governor's position for its "muddleheadedness" and characterized it as "wishy-washy." Various Tocks boosters predicted that Cahill's conditions would be fairly easy to meet. They would be proven wrong. Others sighed in relief since they had expected the governor to oppose the dam outright.[21]

It had been hoped that construction of Tocks Island Dam would have started by 1972. By the end of the year, construction looked more uncertain than ever. Both the Council on Environmental Quality and Congress had told the Corps of Engineers not to proceed. Support was also eroding in the Delaware River Basin Commission, a long-time dam booster. DRBC's approval of construction was not only required by the Delaware River Basin Compact, but it was also the agency that would be paying for the water-supply storage in the dam. Within the commission, New York State had major problems with the eutrophication issue and the contention that it had to do something to reduce nutrients in the proposed reservoir. New Jersey's governor had outlined seven conditions for New Jersey's continued support of Tocks Island Dam. They were not easy conditions to meet.

Thus, for the time being, construction of Tocks Island Dam was at a standstill. What had been seen as so promising only ten years before was now in serious trouble. Would it ever get moving again? Many people hoped not.

11

Pox on Tocks

Public opposition to Tocks Island Dam existed from the beginning in the Minisink Valley. The dam and the recreation area took a lot of land—112 square miles, in fact. This acreage, largely owned by private individuals and organizations, was contained in 7,344 separate tracts of land in twenty-two municipalities in three different states. Although estimates vary, the "wilderness" area coveted by the U.S. government contained 2,400 to 2,600 homes (including seasonal homes), 25 summer camps and sportsmen's clubs, 100 to 125 farms, 100 to 200 nonfarm businesses, and dozens of public buildings, including at least 7 churches and 3 schools. More than 5,000 graves and numerous historical buildings scattered throughout the area testified to the long history of human settlement in the valley.

If you were a landowner in the valley in the early 1960s, Tocks Island Dam and the recreation area might not have made sense. Tocks Island Dam supposedly protected a downstream floodplain of about ten-thousand acres. Because of the added water-supply storage, Tocks Island Reservoir flooded one-third more land than it protected. Forty-seven thousand more acres were wanted for a large eastern national park, and this tripled the amount of land that was to be taken. Few property owners were happy to give up their homes, businesses, and communities so that city folks could have a tax-paid vacation spot a couple of hours away or ample water to waste in their homes and industries.

Among the earliest opponents of the dam was Nancy Shukaitis, a housewife living near the Tocks Island Dam site. Like many Minisinkers, her family had lived in the valley since the 1700s. In the 1940s she and her husband had circulated petitions against one of the New Jersey power-dam proposals, and in the early 1960s she went to Washington several times to oppose the dam and recreation area at congressional hearings. At the March 1965 hearings on the recreation area, she pleaded for a national recreation area based on private land ownership, local land-use control, and federal control limited to technical assistance and oversight functions. Her proposal was ridiculed.

At the March hearings, Mrs. Shukaitis also pleaded for public hearings in the Tocks Island region, where, presumably, Congress would hear more opposition than just her own. Although congressional committees are reluctant to hold hearings away from Washington, Mrs. Shukaitis organized a last-minute campaign that resulted in the House National Parks and Recreation Subcommittee hearing of April 22, 1965, in East Stroudsburg, Pennsylvania. Prior to the hearing, many of the landowners affected by the federal projects formed the Delaware Valley Conservation Association (DVCA) to present a united front at the hearing. Nancy Shukaitis was elected director and chief spokesperson for the organization. For the next five years, the DVCA would be the primary organization fighting Tocks Island Dam.

After the April hearings, the Delaware Valley Conservation Association began to grow. Ultimately, about a thousand persons joined. Locally, the DVCA campaigned against the boosterism of the pro-Tockers. Speeches were given, letters were written, and, when funds were available, anti-Tocks advertisements were placed in local newspapers. In late 1966, for example, one ad's headline read "Hail the Tocks Wreakreation Area!" while another ad (headlined "A Coming Attraction?") showed a photo of mudflats at a Corps reservoir in southwestern Pennsylvania.

The DVCA also lobbied various government officials and politicians. In September 1966 a petition with two-thousand signatures was sent to President Johnson. The DVCA also courted the large national environmental organizations, such as the Sierra Club and Friends of the Earth. Although many of these organizations would later oppose Tocks Island Dam, in 1966 they were still enamored with the idea of the national recreation area surrounding Tocks Island Reservoir.

Early in its history, the DVCA decided to fight Tocks Island Dam and the recreation area in the courts. In the mid-1960s, there was little hope that a tiny organization like the DVCA could halt a major dam project. Their lawyers had told them that. Still, a lawsuit showed the world that the DVCA had "strenuous objections" to the way things were being done in their valley.[1] During the summer of 1966, therefore, funds for the lawsuit were sought from each valley property owner. More than a thousand people contributed, and

Nancy Shukaitis at the 10th anniversary Earth Day activities near Tocks Island, April 20, 1980. The Lenni Lenape chief was a guest for the celebration. (National Park Service Collection)

604 people joined the class-action suit. With its war chest, the DVCA hired a Philadelphia law firm.

The suit of the DVCA and the individual plaintiffs was filed on November 2, 1966, in the United States District Court in Scranton, Pennsylvania. The number of plaintiffs in the suit made it one of the largest ever in the eastern United States. The suit was not against the U.S. government per se, but against the Secretary of the Army, the Secretary of the Interior, and the Corps' chief of engineers. It contended that these three officials were acting unlawfully by proceeding with land acquisition while the Corps was still unsure whether a suitable dam site could be found. The suit also contended that the dam violated federal regulations because the benefit-cost ratio was too low; that the land-acquisition process was unfair because eminent domain was being used for acquiring recreation land; that the park would create a public nuisance due to traffic jams, pollution, crime, garbage, and other effects; that the dam would be unsafe; and that significant historical and archaeological sites would be lost.

The DVCA arguments were never heard in court. Federal attorneys argued that the suit was actually against the U.S. government and not against the three named individuals. If this was true, the government had to agree to the suit before it could be tried. The district court accepted these arguments and dismissed the lawsuit on June 5, 1967. The DVCA appealed the decision to the Third Circuit Court of Appeals in Philadelphia, lost on March 8, 1968, and, finally, appealed to the U.S. Supreme Court. In October 1968 the latter chose not to review the DVCA appeal.

The Delaware Valley Conservation Association continued fighting the dam as the lawsuit dragged through the courts. The major thrust of the DVCA effort continued to be its lobbying and publicity efforts concerning the ills of Tocks Island Dam. Counseling was also provided to people whose land was about to be acquired by the federal government. After the 1965 hearings in East Stroudsburg, members of the DVCA began making two treks a year to Washington to testify against Tocks Island Dam appropriations before the Public Works subcommittees and concerning park matters before the Interior and Parks subcommittees. In 1969 DVCA stories about questionable Corps land dealings led to an investigation by the General Accounting Office. Without a doubt, the DVCA testimony before Congress eventually contributed to erosion of support among some members.

Nancy Shukaitis remained the chief spokesperson of the DVCA during the formative years of the organization. In November 1967 she was elected to the Monroe County, Pennsylvania, Board of Commissioners. Although Tocks Island Dam was not a major issue in the election, her win was seen as a major victory for the local anti-Tocks forces. Beginning with the election of Nancy Shukaitis, each of the four Tocks Island counties would slowly begin abandoning its support for the dam project.

After Nancy Shukaitis became a county commissioner, the leadership of the DVCA passed to various other women. First was Marion W. Masland, a retired schoolteacher from Philadelphia. She was an idealist who had a long association with the Minisink Valley. Her arguments at congressional hearings were well received by some senators. Following Marion Masland as head of the DVCA was Joan Matheson, a Dingmans Ferry housewife, science-fiction writer, and publisher of The Minisink Bull. After 1972 leadership of the DVCA largely fell to Mina Haefele, a New York City museum curator living in the valley, and later to Sandy McDonald, a Sandyston housewife. By this time DVCA's dam fight was closely allied with the efforts of many other antidam organizations.

While the DVCA was going to court for the first time, one of its future presidents returned to the Minisink. Joan Matheson's ancestors had settled in the Minisink in 1746, and Joan had grown up in the valley. Along with her long-time friend Nancy Shukaitis, Joan Matheson would be one of the most formidable opponents of Tocks Island Dam. By coincidence, she was the wife of a Corps of Engineers colonel. In 1959 her husband, David, left the Army, and the family established a home in Dingmans Ferry. Six months after Tocks was authorized, Joan and seven of her neighbors contracted cancer. This she attributed to the stress of the pending land acquisition. Because of the situation, the Mathesons moved to France. They returned in late 1966.

After her return, Joan Matheson became active in the dam fight of the DVCA. Independently, she began publishing an anti-Tocks newspaper, *The Minisink Bull*, patterned after the French satirical weekly *Le Canard Enchaine*. Matheson's publication was an underground newspaper in the sense that it was unconventional, cause-oriented, and antiestablishment. The paper itself was both crude and clever. Its messages were often couched in cute cartoons, euphemisms, double entendres, puns, satire, wordplay, and a Will Rogers style of humor. The primary message of *The Minisink Bull*, however, was crystal clear: The people of the Minisink Valley were being victimized by their government, and the multifaceted Tocks Island Dam Project represented big-government abuse at its worst.

Eventually, *The Minisink Bull* had a four-hundred-copy pressrun. It was published twenty-three times in the period from 1966 to 1970, with each issue displaying essentially the same format: eight pages of "news," editorials, cartoons, and other features. Many people contributed material for the *Bull's* pages, and the paper periodically provided news that was either missed or ignored by the conventional newspapers in the region.

The Delaware Valley Conservation Association, *The Minisink Bull*, and, in some respects, the Lenni Lenape League represented the antidam faction in the Tocks Island region. Over time, their efforts helped change the attitude of the region from pro-Tocks to definitely anti-Tocks. In the process, they raised for the first time many of the environmental issues that would plague Tocks Island Dam after 1970. The local dam opponents helped dispel the mystique of Tocks Island Dam, increased awareness of its many ramifications, and became local contacts when outsiders joined the anti-Tocks fight.

The Tocks Island Dam Project was discovered relatively late by the environmentalists. During the first eight or nine years of the project, the dam had had little environmental opposition save for the Delaware Valley Conservation Association and the Lenni Lenape League. The latter's efforts on behalf of Sunfish Pond, however, had brought widespread attention to the fact that something large and potentially harmful to the environment was happening

Famous anti-Tocks sign by Joan Matheson and Sid Marshall. The humor, wordplay, and message were typical of the *Minisink Bull.* (Albert Dillahunty, National Park Service Collection)

in the Minisink Valley. As a result, various chapters of the Sierra Club and other organizations began taking a hard look at the dam project. After CEQ's critical comments on the Corps' sketchy environmental impact statement were made known, the attention of environmentalists was clearly focused on stopping Tocks Island Dam.

An immediate response to CEQ's comments was the formation of the Save the Delaware Coalition. Organization of the coalition was initiated by Harold A. Lockwood, Jr., a Philadelphia lawyer. In late 1970 Lockwood visited the leaders of the Delaware Valley Conservation Association, the Lenni Lenape League, Nancy Shukaitis, and others to suggest that they and his Eastern Pennsylvania Chapter of the Sierra Club join forces in order to mount a united effort to stop Tocks Island Dam.[2] Within five months, the coalition had enlisted twenty-eight member organizations to its cause. It grew to more than fifty members by 1972 and to more than sixty by 1975. The organizations within the Save the Delaware Coalition represented thousands of people; thus, the coalition was a potent political opponent of the dam.

The Save the Delaware Coalition was a loose confederation of diverse organizations united in their opposition to Tocks Island Dam. National members included the Sierra Club, Trout Unlimited, the Wilderness Society, the National Wildlife Federation, and others. Local members ranged from the Essex County Young Republicans Club to thirteen chapters of the Sierra Club, chapters of other national organizations, various sportsmen's clubs, and more. Most member groups had their own particular concerns. The sportsmen's clubs were interested in the impact on fish, the Daughters of the American Revolution in the drowning of historical sites, the American Canoe Club in the loss of the free-flowing river, and so on. The DVCA, a charter member of the coalition, was opposed to the recreation area as well as to the dam, a view not shared by the other members. Some member groups were very active in the dam fight and not only supported the Save the Delaware Coalition but launched their own anti-Tocks crusades as well.

The chairman of the coalition from the beginning was Harold Lockwood. In some respects, members of the Pennsylvania and New Jersey Sierra Club chapters dominated the organization. The coalition presented a united front of dam opponents and provided a network to apprise each member group of Tocks-related activities. On the national and state levels, Lockwood and others conducted intensive lobbying efforts on behalf of the organization. These and other activities made the average citizen (and politician) aware of the negative side of Tocks Island Dam.

The environmental opposition to Tocks Island Dam involved many issues. Mudflats, eutrophication, the loss of fish and wildlife, reservoir pollution, and the loss of the free-flowing river were issues that persisted throughout the controversy. Other issues also flared up briefly and then became less important. For example, questions about dam safety were first raised by the Delaware Valley Conservation Association. Later this issue was transformed into an emotional fear of dam failure from earthquakes.[3] Almost anybody could find something in the dam project to be concerned about.

The link between Tocks Island Dam and water needed for nuclear power provided another emotional issue: Was the dam being built for the people or for the large public utility companies? The specter of nuclear power plants using Tocks water served to reaffirm this idea in many minds, even though the Delaware River Basin Commission was almost as concerned about the issue as anybody. The only state or federal agency to officially oppose the dam, however, was the Pennsylvania Fish Commission, which alleged that the dam would cause a loss of river fishing and severe reductions in the annual shad runs.[4]

During the controversy, some dam opponents tried to tie the construction of Tocks Island Dam to New York City's use of the Delaware. The city appears to have had only a peripheral relationship to the dam, although the reservoir would have drowned the Montague gage and reduced the basin's reliance on the flow releases of its reservoirs. This did not stop some opponents from claiming that Tocks planners were secretly planning to give the city Tocks water. Once New York City was mentioned, the long-standing complaints about the city's leaky water system, as well as other anti–New York City complaints, were raised. Many of these had been around since the first Supreme Court fight.

Lower-basin people had their own issues. Gretchen Leahy and her Pollution Control Group of Lower Bucks County were most concerned about the water needs of the upper Delaware Estuary and the impact of Tocks Island Dam on its water quality.[5] They contended that DRBC's upper estuary sewage-treatment requirements were too high, and in fact had been designed to offset the pollution loads that would be flushed out of Tocks Island Reservoir (dead organic matter, suspended solids, and other material). When DRBC instituted basinwide water charges in the early 1970s to pay for the water-supply storage in the Corps' reservoirs, the group challenged the regulations in court. They argued that the Delaware Estuary community was being forced to pay twice: for river water that would not reach the estuary since it was going elsewhere (to New York City, North Jersey, and power plants); and for the new sewage-treatment plants that would not be needed if a free-flowing river was maintained upstream of the Delaware Estuary. They believed that DRBC was controlled by out-of-basin (i.e., New York) interests.

The major thrust of the anti-Tocks effort was to stop the dam (for whatever reason was given) but to keep the recreation area. Typical of environmental controversies, a preservationist viewpoint existed. Most Tocks opponents wanted a wilderness-like park around a damless, natural river. However, throughout the dam fight, nobody was sure if the DWGNRA could legally exist without the dam. Its legislation linked it to the construction of Tocks Island Dam, and there were some other major issues as well.

The first party to raise questions about the separability of the dam and recreation area was the Corps of Engineers.[6] The National Environmental Policy Act had required that a "no action" alternative—i.e., the option of doing nothing—be addressed in environmental impact statements. In addressing this option, the Corps questioned whether scrapping Tocks Island Dam would lead to the deauthorization of the DWGNRA. It was a question that no one, including the Council on Environmental Quality or the National Park Service, could answer definitively.

Even before that question could be answered, another issue had to be addressed. Was a river-based recreation area without the drawing power of a large lake even feasible? Would it have valid and useful purposes? CEQ asked the National Park Service to answer these questions, and during the summer of 1971 the agency developed A Natural System Plan for the Delaware Water Gap National Recreation Area. The study was not made public until the Sierra Club and the Save the Delaware Coalition obtained copies and released it to the press in September. The park service's reluctance to release the report reflected its continued interest in a large-scale, reservoir-oriented recreation area. Within the agency, however, there was a minority that considered a river-based recreation area to be an immensely attractive idea.[7]

The National Park Service's quick study indicated that a recreation area built around a free-flowing river was a viable option. The "natural systems" plan deemphasized reservoir recreation and promoted recreational pursuits like canoeing, river swimming, hiking, winter sports, environmental education, and other outdoor activities. The results of the park service study delighted the environmentalists. Here was justification for their dam opposition.

The leaders of the antidam forces realized that more than just the value of a river-based park had to be proven. Alternative ways of obtaining the water-supply, flood-control, and power benefits of Tocks Island Dam had to be found. It was clear to them that no government agency was going to search for these alternatives as long as Tocks Island Dam was still on the books.

The technical analysis of the Tocks Island Dam Project was encouraged from the beginning by Harold Lockwood and others.[8] Because CEQ and Congress had stopped dam construction, albeit temporarily, time was available to conduct such studies. The delay also meant that the environmentalists did not have to spend energy and resources on a legal challenge. Instead, Tocks Island Dam became one of the few environmental controversies decided outside the courtroom. The idea of challenging the dam project on technical grounds appealed to many Tocks opponents who believed that the dam was vulnerable once the facts were known.

Among the first technical analyses by the environmentalists was one done by a trio of Princeton scientists: Smith Freeman, Edwin S. Mills, and David Kinsman. In 1971 the three conducted a quick analysis of high-flow skimming as a potential way of getting the 300 mgd that New Jersey wanted from Tocks Island Dam. High-flow skimming is simply the withdrawal of river water during high-flow periods for use during low-flow periods. In the intervening months, the water is kept in off-stream reservoirs. Freeman, Mills, and Kinsman concluded that high-flow skimming was not only feasible but much cheaper than getting water from Tocks Island Reservoir.

In 1973 the Environmental Defense Fund (EDF) followed up on the Princeton work and hired an engineering firm to examine Tocks water-supply alternatives using more conservative design standards. M. Disko Associates completed its study in October. The study, *New Jersey Water Supply: Alternatives to Tocks Island Reservoir*, investigated various water-supply alternatives but concentrated on high-flow skimming. The consultants revived the decades-old idea of using the Flat Brook Valley as the location of an off-stream storage reservoir. Although the idea was feasible, the proposal to dam one of New Jersey's premier trout streams was not at all palatable to the environmental cause.

The Environmental Defense Fund's first foray into the Tocks Island controversy had actually been a critique of the dam's costs and benefits. The study, released in early 1972, concluded that costs were underestimated and that benefits were overstated. Of particular interest, however, was the conclusion that dam construction might encourage development on the floodplains below the dam. EDF's Long Island office then initiated an intensive land-use inventory of the floodplains as far south as Burlington, New Jersey, on the upper Delaware Estuary. The purpose of the study was to learn what the effects of a flood like that of 1955 would be if it occurred in the 1970s. If development on the floodplains was less now than it was in 1955, the flood-control benefits of Tocks Island Dam were overstated.

The products of EDF's flood-control work were two reports: *Flood Control—A Field Investigation*, published in September 1972, and *Flood Control and the Delaware River*, published in April 1973. The studies concluded that the number of structures located on the Delaware River floodplain had declined drastically since 1955. Flood-control works had been constructed, old buildings had been razed, riverfront parks had been created, and new buildings were being floodproofed. Many of these improvements were the direct result of federal programs like those for flood insurance and urban renewal. EDF promoted this nonstructural approach to flood control as an alternative to Tocks Island Dam and called on the Corps and the Delaware River Basin Commission to institute such a program instead of relying on flood protection from dams.

In 1973 the Save the Delaware Coalition published its own ambitious report, entitled *The Tocks Island Dam, A Preliminary Review: Papers in Support of a Free-Flowing Delaware River.* The 190-page book was purportedly "just the beginning of the analysis to tell the full story of the Tocks proposal."[9] The report did in fact present a multifaceted look at the dam project and the alternatives to its construction. The coalition's book was actually a collection of papers and other material that had been edited by two Bell Lab employees, George E. Schindler and Frank W. Sinden, into a relatively cohesive, technical argument against dam construction. The report presented few traditional environmental arguments against dam construction. Instead, it attempted to show that there were other ways of providing flood-control, water-supply, recreation, and power benefits than building Tocks Island Dam.

A much less "preliminary" report was soon to come from the Save the Delaware Coalition. For some time the National Park Service's natural-systems plan had been gathering dust. As long as the plan was ignored, lake-based recreation would continue to be a forceful argument for building Tocks Island Dam. Logically, the next step in the planning process was the drafting of a master, or sketch, plan. No one in the National Park Service had been assigned this task. Sympathizers on congressional staffs then suggested that the coalition develop a river-based recreation area plan.[10] Coalition members were very receptive to this suggestion since it represented a positive side to their dam opposition. With funds raised from its members, the organization hired the consulting firm of Caneub, Fleissig and Associates to develop the plan. In the spring of 1974, the firm published *A Concept Plan for the Delaware River Park.* It was one of the highlights of the environmental opposition to Tocks Island Dam.

The Save the Delaware Coalition plan began with the National Park Service's natural-systems plan and incorporated ideas that had been promoted by various dam opponents. Frank Sinden, a member of the North Jersey Chapter of the Sierra Club, had, for example, been very interested in a park that was not automobile-oriented. Others had their own ideas to offer. The Concept Plan thus became a forum for presenting and analyzing various visions of a Delaware River park. Overriding all the personal ideas was the feeling that the unique natural beauty, high environmental quality, and rich history of the Minisink Valley offered much more to the public than a man-made lake could.

The Concept Plan presented an exciting but low-key linear recreation area that supposedly offered visitors a rich and diverse outdoor experience. Unlike the park envisioned by the Tocks proponents, the river-oriented park had almost nineteen-thousand acres of relatively flat floodplains that could be used. This land would be flooded by the reservoir if the dam was built.

The idea of a river-oriented DWGNRA practically sold itself. Neither expensive recreational facilities nor dam was needed. The natural beauty, the historical and cultural features, the wildlife, the topography, the small lakes, the waterfalls, the fish were all there; so was the Delaware River, one of the last major free-flowing rivers in the country. Photos taken by dam opponents and concise descriptions of the area's beauty were added to show what the Minisink Valley and the Delaware River looked like in the absence of a dam. The coalition wanted to make sure that decision makers in Washington, D.C., and elsewhere knew what was being offered. The WRA/DRB had done much the same thing in much the same way in its DWGNRA promotion brochures of the early 1960s.

Two alternative plans were presented in the Concept Plan. The first was automobile-oriented, with numerous access points up and down the valley. The second alternative, which was the more favored plan, was largely automobile-free. It envisioned major activity nodes at the northern and southern ends of the park, with secondary activity nodes at Bushkill and Dingmans Ferry in Pennsylvania. The nodes served as access points, with provisions made for parking, food, and other services. From the activity nodes, visitors were to be transported into the park by specially equipped electric buses. Riders would be permitted to get off at any point along the river; thus, park visitors would be distributed over a widespread area to pursue their own individual activities. Also serving as part of the internal transportation network were to be numerous hiking and bicycling trails.

In many respects, the variety of recreational activities provided for in the Concept Plan was not much different from the activities presented in the DWGNRA Master Plan. Besides the obvious difference between a reservoir and a free-flowing river, the major difference was the number of annual visitors. The Concept Plan reduced the number of visitors to four million per year, not by coincidence the number of visitors desired by New Jersey's Governor Cahill. The report also emphasized that these four million visitors were distributed more evenly throughout the year and that the retention of floodplains meant that visitors had an additional nineteen-thousand acres to get "lost" on. In any case, the Concept Plan proposed controlling visitation by monitoring use and instituting a system of permits and reservations. The Concept Plan was a bold, albeit idealistic, vision of what a national recreation area should look like.

Most of the various reports prepared by the environmentalists were scrutinized by the staff of the Delaware River Basin Commission, the Interior Department, the Corps, and other agencies as applicable. Reports that suggested alternatives to Tocks Island Dam were given special attention. The Tocks Island Dam proponents generally felt that the voluminous *Delaware River Basin Report* had examined and rejected so many alternatives that no more analyses were needed. Alternatives were always considered impractical for various reasons. High-flow skimming, for example, was discounted because it required reservoirs to be built in Pennsylvania and New Jersey. Who was going to build these nonfederal reservoirs, and what would their impacts be? Desalinization, a technique for making estuary water fit for drinking, was seen as being too expensive and energy-intensive.

Nonstructural flood-control measures were favored by the water agencies, but not as the sole solution to flood protection. The Delaware River Basin Commission, for example, was already promoting elements of nonstructural flood control by 1970.[11] Alternatives to pumped storage, such as the use of gas-fired turbines, were considered expensive or undependable. A river-based recreation area was considered elitist and somehow seemed to discriminate against the urban masses who were planning to flock to the reservoir. Water-supply alternatives that involved dams on tributaries were rejected because they also had environmental impacts. When dam opponents brought up New York City issues, dam proponents countered that the city's diversion was fixed by the U.S. Supreme Court and the Delaware River Basin Compact and had nothing to do with Tocks Island anyway.

Thus it went: The Tocks defenders had an equally persuasive argument for rejecting every proposed alternative. The technical battleground would decide nothing. The real argument was ideological: Either you believed that Tocks Island Dam was the long-awaited answer to the water needs of the Delaware River Basin, or you didn't. It was like religion.

The pro-Tocks arguments were spelled out many times during the controversy. Most were restatements of the need for the project, with some attempt to counter whatever environmental argument was then in vogue. The alleged benefits of the project—flood control, water supply, power, and recreation—were spelled out again and again. In the fall of 1971, for example, the WRA/DRB published a fifteen-page booklet entitled *The Keystone Project, Tocks Island Revisited.* The booklet revisited the project benefits in detail, updated the dam's status, and then asked and answered thirteen "environmental" questions. The booklet was widely distributed but had no effect on the rising public opposition to the project.

A much more detailed pro-Tocks report was issued in May 1971 by four pro-Tocks congressmen. The purpose of the report was clearly to head off problems that were emerging in Congress. In it, the Corps, the Interior Department, the Delaware River Basin Commission, the Federal Power Commission, the Tocks Island Regional Advisory Council, and the Office of Management and Budget attempted to answer eighty-three questions concerning the dam, the recreation area, and the Kittatinny Mountain Pumped-

Storage Project. The questions covered the whole gamut of Tocks issues. The lengthy document was distributed to Congress, published in the *Congressional Record*, and made available to the news media and the general public.

Also issued in 1971 were various papers concerning aspects of Tocks Island Dam. *Flood Damages and Tocks Protection* and *Water Demands in the Delaware River Basin as Related to the Tocks Island Reservoir Project* were two such reports. These were prepared by the Corps and the Delaware River Basin Commission respectively. The latter report showed rapidly increasing water-supply needs through the year 2020 and concluded that "there is no dependable alternative to Tocks Island Reservoir for meeting the projected demands for water to replace the water to be consumed in and exported from the Delaware River Basin."[12] The paper rejected the high-flow skimming proposal of the environmentalists as too costly.

Mother Nature gave Tocks Island Dam a major boost in 1972 when Tropical Storm Agnes wreaked havoc in the neighboring Susquehanna River Basin. Flood damage was widespread, and several flood-related deaths occurred. Among those directly affected was Pennsylvania's Governor Milton Shapp, a major supporter of Tocks Island Dam. Shapp was forced to evacuate the Governor's Mansion when the floodwaters entered the mansion's first floor. Although most of the Delaware River Basin was spared Agnes, flooding did occur in the western portion of the Delaware River Basin.

The Corps would claim that if Tropical Storm Agnes had taken a more easterly route the Delaware River would have crested five feet above the level of the devastating 1955 flood. Damages due to this hypothetical flood were estimated to be $190 million, but 80 percent of the damages were supposed to be preventable if Tocks Island Dam were built.[13] In the following year, the Environmental Defense Fund took great pains to point out that the Corps' quick post-Agnes analysis greatly oversimplified things.[14]

Nongovernment entities also added to the pro-Tocks arguments. Most of these fact-finding studies were merely reiterations of government reports and pro-Tocks arguments. The WRA/DRB, the Delaware Valley Council, and the New Jersey State Chamber of Commerce were notable examples of organizations that attempted to add their weight to the pro-Tocks side of the controversy by publishing "fact-finding" reports.

The purpose of the publicity efforts on both sides of the controversy was not to convince the other side but to neutralize it. Both sides in the controversy recognized that the power to decide the fate of Tocks Island Dam rested with Congress and the Delaware River Basin Commission. Both Congress and the commission, however, were hung up on the various CEQ concerns and the seven conditions of New Jersey's Governor Cahill. Dam supporters realized that if these concerns were resolved construction of Tocks Island Dam would begin.

Throughout the controversy, CEQ's major problem with Tocks Island Dam centered on the eutrophication issue. The many biological and technical studies, however, never yielded definitive answers about the potential for eutrophication or the magnitude of the problem if it did occur. In lieu of definitive answers, CEQ pressed for assurances that a nutrient-control program would be implemented in the drainage area above Tocks Island Reservoir. New York State was unwilling to implement such a program for its portion of the basin. The Delaware River Basin Commission had the power to institute such a program, but it did not have the resources.

All this changed somewhat when Congress passed the Water Pollution Control Amendments of 1972 (P.L. 92-500), a new federal approach to water-pollution control. During the debates on the bill, Representative Frank Thompson inquired whether the bill would address problems like the potential eutrophication of Tocks Island Reservoir. The House Floor manager, Representative Robert E. Jones of Alabama, assured Thompson that the act "contemplates the control of the water quality problems of the authorized Tocks Island Reservoir."[15]

After the federal pollution bill was enacted in October 1972, dam proponents argued that New York's assurances were no longer needed. The new act required all sources of pollution to be cleaned up by 1983 and eliminated by 1985. The act not only addressed traditional sources of pollution, such as sewage and industrial discharges, but also nonpoint sources, diffuse sources of pollution generally caused by runoff from precipitation. The chicken and cow wastes in the rural New York portion of the Delaware River Basin were prime examples of nonpoint pollution.

In late 1972 and early 1973, the Delaware River Basin Commission took action to counter CEQ's and Governor Cahill's objections to the dam. First, the commission amended its Tocks Island pollution-control plan to state that it would "take all steps to control point and non-point sources of pollution to the extent that they may be shown by field investigations and analysis to contribute to the harmful eutrophication in the Tocks Island Reservoir."[16] In May 1973 DRBC further modified its policies by adding a statement that nonpoint sources were to be managed in accordance with the policies being developed at the federal level by the U.S. Environmental Protection Agency (EPA).[17] This action was an attempt to divorce the nutrient-control issue from the dam controversy by passing the buck to the EPA and the states.

At its May 1973 meeting, DRBC also modified its Tocks Island pollution-control plan to let New Jersey build the subregional sewage-treatment plants desired by Governor Cahill. At the same meeting, DRBC also took care of another Cahill concern by adopting a resolution that limited the DWGNRA's design capacity to forty-two thousand persons per day.[18] This translated to the annual visitor load of four million that Cahill had requested. The change was a paper change that took into account the phased development of the park. Down the road, the recreation planners were still assuming that the DWGNRA would be getting more than ten million visitors per year. It was hoped that the limits on visitors would take care of another Cahill condition: New Jersey would not have an immediate need for new highways.

After the May 1973 DRBC meeting, Governor Cahill acknowledged that two of his seven objections had been met.[19] He was still dissatisfied, though. Cahill continued to look to the federal government for highway funding and compensation for the loss of local tax revenues, to New York for assurances that it would control nutrient runoff, to Pennsylvania for the passage of flood-control legislation similar to his state's new law, and to his own state for passage of a land-use control bill for the Tocks Island region.

The Delaware River Basin Commission actions of May 1973 had also failed to resolve CEQ's objections. CEQ still demanded that New York State agree to implement a waste-management program for the upper Delaware. It specifically wanted to see a program that included 95 percent phosphate removal from municipal effluents and the cleanup of agricultural wastes. Some issues would not go away.

In 1973 Tocks supporters had hoped that their actions and arguments would break the Tocks logjam in Congress. Support for the project was marshalled, and more than seventy-five individuals and organizations presented testimony favoring dam construction during congressional appropriation hearings.[20] The final funding vote, however, was influenced by CEQ's and Cahill's outstanding concerns. Both the House and the Senate approved funding for dam construction but again stipulated that it could not be used until the eutrophication, nutrient-control, and highway issues were resolved. The committee, however, believed that the actions of the project sponsors had taken care of most of the questions.[21]

New Jersey's position on Tocks Island Dam became clouded during the summer of 1973 when Governor Cahill lost the Republican primary in his bid for reelection. In January 1974 the new governor took office. Governor Brendan Byrne, a Democrat, was not bound by the Tocks Island concerns of the Cahill administration. Byrne was not the only new face on the Delaware River Basin Commission in 1974. In New York, Malcolm Wilson had succeeded Governor Nelson A. Rockefeller, the last original member of DRBC, and in the previous year Delaware's governor, Russell W. Peterson, had been replaced by Sherman Tribbitt. Peterson was subsequently appointed, some-

what ironically, to head the Council on Environmental Quality. The potential ramifications of these changes on the fate of Tocks Island Dam were not
known as the project entered its twelfth year.

After several years of heated controversy and delay, 1974 started out looking like the year that construction of Tocks Island Dam might begin. Due out
early in the year was a supplement to the Corps of Engineers' environmental
impact statement. It was hoped that this document would answer the critics
of the dam and the concerns of CEQ and New Jersey. Once these concerns
were satisfied, the Delaware River Basin Commission could bless the project
and Congress would release the construction money that it had already appropriated. There were few optimists, however.

In early May the Byrne administration's new commissioner of environmental protection took office. David J. Bardin had recently returned from four
years of service with Israel's environmental protection agency. On the top of
his list of things to do was Tocks Island Dam. A task force headed by Thomas
P. O'Neill was created to again reevaluate New Jersey's position on Tocks
Island Dam. Until the reevaluation was completed, Governor Byrne was off
the hook.

Meanwhile, New York's new governor, Malcolm Wilson, had taken a firm
position on Tocks Island Dam. On May 2 he sent a letter to Joe Evins, the
chairman of the House Public Works Subcommittee, telling him that New
York was against the construction of Tocks Island Dam at that time. Wilson
wanted completion of the Corps' environmental impact statement process
plus "satisfactory assurances that New York municipalities or private individuals or industry upstream will never be required to spend their own funds to
achieve a level of treatment beyond which could be reasonably expected if
the reservoir was not present."[22] By demanding these assurances, New York
was turning the tables on CEQ.

By this time Congress had become impatient with the Delaware River
Basin Commission's indecision on the dam project. Congress could overlook
CEQ's objections since CEQ was an executive-branch agency without any
legal power. DRBC was another matter entirely. The Delaware River Basin
Compact required all water-resources projects to be approved by DRBC.
More important, the commission represented the views of the four Delaware
River Basin states and was to be the Corps' local sponsor. With DRBC's
support, Congress had invested $54 million on a Delaware River dam. It
needed to know whether to keep spending money on the dam or to abandon
it. With proponents and opponents about evenly matched, however, DRBC's
indecision promised to drag on forever.

There were two ways to get the Tocks Island Dam Project off dead center.
The first was to allow the project to move forward and resolve the environ-

Anti-dam protesters accompany New Jersey Governor Byrne on his 1974 canoe trip
through the Minisink Valley. (Ray Fauber, National Park Service Collection)

mental questions in the ensuing court case. Opponents of the dam had
promised legal action over the project's compliance with the National Environmental Policy Act. The legal challenge, however, could not take place
unless Congress gave the Corps an unrestricted green light on construction.
Among the persons favoring the judicial solution was Representative Frank
Thompson, the dam's staunchest proponent in Congress. According to
Thompson, "It seems to me that it would be most appropriate at this
time . . . [for the Corps] to advertise for bids and, if CEQ or anybody else
feels that they have got a case against the project, let them for heaven's sake
take it to court and get this issue resolved once and for all."[23] Others shared
Thompson's feelings, including Joe Evins, the powerful chairman of the
House Public Works Subcommittee. Evins had little sympathy for environmental causes.

The other method for getting the dam controversy resolved was to have an
unbiased third party conduct a study of the project. CEQ had proposed such a
study in 1971, suggesting that the National Academy of Sciences would be a
good candidate for the job. The idea was subsequently dropped, but it was
never forgotten. In early 1974 Russell Peterson, the new head of CEQ,

suggested it again. Senator Clifford Case of New Jersey, who had gone from a dam supporter to a dam opponent, also favored such a study. By June Case and six of the Delaware River Basin senators were calling for an impartial study of Tocks Island Dam.

The two proposals for ending the Tocks Island logjam were debated. There was resistance to more studies, particularly in Evins's House Public Works Subcommittee. His committee therefore voted to give the project unrestricted funds with which to begin construction. Opposing this move was Representative duPont of Delaware, who, with others, threatened to mount a floor fight if the Appropriations Committee endorsed the Public Works Subcommittee recommendations. Eventually, the House agreed to the full Tocks budget request but again stipulated that the money could be used only for land acquisition.

The Senate was even more adamant that the Tocks Island Dam question should be resolved. A joint House-Senate Conference Committee therefore decided that $1.5 million should be allocated for a comprehensive review study. The Senate, however, did not want the dam project receiving any money other than the $1.5 million. As a result, for the first time no money was allocated for land acquisition. This action forced the Corps to transfer $3 million from other projects in order to meet its legal commitments in the Tocks Island region.

One issue, the question of who would manage the study, had to be resolved before Congress completed its work. The extremely tight schedule precluded the use of the National Academy of Sciences. There was sentiment for having the Delaware River Basin Commission perform the study, but in the end the Corps was given the task. The environmentalists opposed this, claiming it was akin to allowing the "Tocks Fox" to guard the henhouse.[24] Congress, however, did not want to cast aspersions on the Corps' past work by selecting someone else. As a partial compromise, Congress required the Corps to conduct the study in cooperation with the Delaware River Basin Commission. The commission, in turn, was to conduct an unprecedented public-information program during the study.

Congress wanted more than just another study. It wanted the Delaware River Basin Commission to resolve the dam controversy. DRBC was given one year from the date that Congress appropriated the review study funds to decide whether Tocks Island Dam should be built. On August 28, 1974, the review study funds were appropriated. The clock was now ticking for Tocks Island Dam.

12

Troubles in the Minisink

If there was any doubt that Tocks Island Dam had its negative side, one only needed to look at the Minisink Valley in 1974. During the four years that the dam's construction had been debated, the pastoral valley had become a war zone. The land-acquisition process, one of the largest of its kind, had created some major problems.

Land acquisition for the dam and recreation area projects began in 1967. The process was supposed to be orderly and fair. Acquisition was to start in the Delaware Water Gap and proceed through the valley. In the sixth and final year, land in the Milford, Pennsylvania, area was to be acquired just as Tocks Island Dam began backing up water.

The mechanics of land acquisition were seemingly simple, too. After some preliminary work, Corps appraisers would visit a property and develop an estimate of its fair market value. After this, Corps negotiators were to talk to the property owner. There were two possible outcomes from the negotiations. If a selling price was agreed upon, the government purchase proceeded like a normal real-estate transaction. If agreement could not be reached, or if title problems existed, the government instituted condemnation proceedings and the case went to federal court. Condemnation only established the purchase price. The government still took possession of the property.

During the selling process, the property owner had several options: He could elect to live in his house for one year after purchase; he could receive salvage rights to move his house to some other location; or, if the property was in the DWGNRA, he could take a life tenancy option, which allowed him to live in his home for up to twenty-five years. Each option was at the discretion of the federal government, and each had certain penalties associated with it.

Questions concerning the fairness of the process plagued it from the beginning. Even before it began, there were concerns that the National Park Service had drawn the final DWGNRA boundaries in such a manner as to exclude certain influential persons and businesses. The jagged park boundary had no rhyme or reason, and whose property was in or out was largely a discretionary decision.

Suspicions were thus heightened as the first purchases were made. The early purchases did nothing to dispel these suspicions, either. The price paid for an individual property appeared to be influenced by who owned it and not by its comparable worth. Owners of billboards situated on small tracts of land, for example, were paid well when compared to various small businesses that were purchased at the same time. The small businessmen also complained about problems in dealing with the Corps' negotiators, and each experienced drastic losses in annual income due to the loss of property.[1]

The early complaints about the purchase price reflected an issue that would plague the land-acquisition process. Many property owners believed that the government did not pay fair market value for their property or that the price paid did not allow purchase of equivalent property outside the park boundaries. A partial explanation for these complaints was that pending federal acquisition depressed property values since less property exchanged hands. Meanwhile, land prices were rising outside the area because of inflation. Adding to the woes were rising mortgage interest rates, which added to the replacement costs for many persons.

A more damaging explanation for the complaints exists. Did the government try to buy property for the cheapest possible price? The answer is probably yes. The Corps' initial offer on an individual property was determined by the appraised value and "such factors as the property owners' age, business acumen, and whether the owners were represented by counsel."[2] These factors had nothing to do with fairness. The government, of course, was under no obligation to pay the highest possible price or to pay replacement costs. With the original cost of land acquisition too low and federal dollars being diverted to Vietnam, funds had to be stretched as far as possible.

On the other side were complaints that so-called land speculators made out too well. Some appear to have been opportunists who took advantage of the chaotic situation by buying land at low cost from people who had to sell before the government got around to buying their land. There were probably others, too, who had the business acumen, legal counsel, and possibly the political clout to extract the best deal possible from the government negotiators.

In other cases, people trying to cooperate with the government were sometimes left high and dry financially when the scheduled purchase of their property was delayed. In these cases, revenue-producing livestock was sold or scarce funds were invested in new homesites. Another problem was the variety of persons that the government had to deal with. Some had special circumstances that could not be handled by inflexible government purchasing policies. Others had never traveled outside the area, had never bought or sold real estate, had no idea of what their property was worth, and did not understand the finer points of the whole process. Good and bad Corps negotiators also existed. Many persons were treated kindly, but some people complained about high-pressure tactics, harassment, and duplicity. It was a mess.

The ultimate option for a property owner was to go through the condemnation process. This option had its own set of problems. Going to court jeopardized the potential selling price of the property. Not only might the jury set a price that did not justify the expense of hiring lawyers and private appraisers, but only the initial price offered during the negotiations was admissible in court. This meant that property owners could conceivably get less for their property than the final offer made during negotiations. Thus, going to court was not only an expensive hassle but also risky.

Going to court was not a viable alternative for elderly people who could not travel, for people tied to jobs, or for others in special circumstances. If a property owner did go to court, arrayed against him would be federal attorneys interested in advancing their careers. Although condemnation was supposed to determine just compensation in a noncompetitive trial situation, this was not always the case.[3]

As more and more land fell into government hands, new problems arose. The area became increasingly desolate, with declining road care, police, and fire protection. As a result, many of the newly emptied houses were occupied by squatters or picked over by scavengers and vandalized. Arson became a widespread fear. Between June 1970 and June 1972, for example, about three dozen buildings in the valley were burned. Trespassing also increased because it was no longer possible to tell government land from private property. Adding to the fear were several cases in which government demolition teams accidentally destroyed the wrong house. The remaining residents, many of them elderly, became afraid to leave their homes for fear that something would happen in their absence. Many no longer had nearby neighbors.

Other problems existed, too. Many were as individual as the persons being affected. Special situations existed all over the valley, but the government regulations were applied uniformly. Resentment resulted when people realized that they were losing their homes so that city folk could romp on their former land or drink water from a reservoir covering their former gardens. Many of these properties had been in the same family for many generations or represented the culmination of a life-long dream. When park service families moved into the newly acquired houses or when a group of commercial artists was brought in to occupy a former village (as in the case of the Peters Craft Village), the resentment increased.

The most serious problem may have been psychological. Moving is never easy; it is recognized as one of the most stressful undertakings a person can go through. Being forced to move is even worse, especially if accompanied by a chaotic acquisition process or the feeling of being cheated. A general impression exists that the stress of the Tocks Island process resulted in the premature death of many elderly persons. At least one suicide resulted from the land-acquisition process, and some residents left the area with a bitterness that has prevented them from ever returning.

The problems in the Minisink coincided with the debates over Tocks Island Dam. The plight of the local residents generated a great deal of sympathetic press coverage. The Delaware Valley Conservation Association, serving as the local arm of the antidam forces, made sure that the horror stories in the valley were told. The association sponsored press bus tours of the scarred areas, and during the tours guides like Mina Haefele bitterly told of the problems they and their neighbors had had with the government. The stories did not flatter the federal agencies operating in the Tocks Island region. Today, ten to twenty years later, much of the bitterness remains.

One of the land-acquisition problems in the Minisink Valley evolved into a bizarre anti-Tocks effort. Causing major headaches for both the local residents and the Corps were squatters who occupied empty houses in the lower part of the valley. The core of the squatters (or river people) was made up of 1960s-style hippies who illegally occupied the valley for five years. Their dream was different from the visions of the pro-Tockers or the wilderness vision of the environmentalists. The hippies wanted to establish a counterculture, self-sufficient agrarian society based on the rich soils of the Minisink. As a result, they contested the federal government's right to build Tocks Island Dam.

The squatter problem was created by the Corps of Engineers. By mid-1969, the Corps had accumulated a collection of empty houses and other buildings near the Tocks Island Dam site. The Corps offered these buildings for rent by advertising in various newspapers, including New York City's *Village Voice*. The leases were supposed to be temporary since the Corps believed that dam construction would soon start. The rental income was to help offset the loss of local tax revenues, and it was hoped that renting would reduce the vandalism in the area.

Among the persons answering the Corps' advertisements was a group of hippies living in New York City's Lower East Side. Like most hippies, they wanted to create a new society based on love, brotherhood, soft technology, and a free-spirited life-style. A piece of land containing a large farmhouse, numerous cottages, and fertile river bottomland was leased about one mile upstream from the Tocks Island Dam site in New Jersey. The farm-commune was named "Cloud Farm," and its members moved to the valley to establish an alternative life-style. Friends joined the Cloud Farmers until all the houses on the New Jersey side of the river were occupied. Most of the hippies were actors or artists of one kind or another.

By the summer of 1970, there were no more houses vacant on the New Jersey side of the river. The Corps, meanwhile, had acquired property across the Delaware River in Pennsylvania, and the hippies began moving there. The first house occupied in Pennsylvania was a large farmhouse on the Delaware River upstream from Tocks Island. It was occupied by Bill Read, a friend of the Cloud Farmers, and his family. The Reads were soon joined in the farmhouse by Dorothy Belmont and her husband, former neighbors of the Reads in New York, and by another couple. Soon, by word of mouth and largely unknown to the Corps, more hippies learned of the opportunity and also moved into the valley.

In 1970 the Corps allowed the leases on the New Jersey properties to expire. Many of the houses, however, had no lease, and rent was rarely paid on those that did. The Cloud Farmers and other New Jersey hippies were then evicted by the New Jersey State Police, and the dwellings were demolished. Meanwhile, squatters had completely taken over the lower Minisink Valley in Pennsylvania.

The territory of the Pennsylvania squatters was a five-mile section of River Road beginning several miles north of Shawnee-on-the-Delaware and running to near Wallpack Bend. About two dozen houses and the historic Zion Lutheran Church were occupied by various families and groups of individuals. The church, located on a hill overlooking the valley, was renamed the "Church of Ecology," and marijuana grew in its cemetery. Scattered among the houses were tent camps, Indian teepees, homemade structures, and a variety of innovative hippie homes, including one geodesic dome built on a raft in the Delaware River.

During the summers of 1971, 1972, and 1973, the permanent, year-round squatters were joined by hundreds of others. Most were simply youths looking for a "scene" to be part of, but the transients had their share of bad apples,

Squatter musicians at the August 1971 Squatters Fair. The fair featured music, art, and homegrown produce. (Albert Dillahunty, National Park Service Collection)

too. The *Pocono Record,* in a series of articles in 1972, found much to criticize about the Minisink hippie scene. It described the scene as "a major station on the underground hitchhikers highway linking New York with Canada and Florida" and cited numerous incidents in the valley.[4] Even the *Pocono Record,* however, recognized that two classes of squatters existed in the valley: "There are people along the Delaware whose motives in building for themselves a different, but good life can not be imputed. But they are there illegally. And because they are, they attract others who have caused a massive (for this area) increase in crime and the potential for far worse."[5]

The year-round squatters (hereafter the Squatters) were opponents of Tocks Island Dam and true believers in the hippie philosophy. Individuals came and went in the group, but many lived in the valley from 1970 to 1974. About four or five dozen people were part of the group, and it eventually became a rather cohesive community of free-spirited young adults. A number were artists, musicians, or craftsmen. Some had young children, and four babies were born in the valley during the Squatters' stay.

The back-to-the-land interest of the Squatters led to the establishment of gardens and the raising of chickens, goats, rabbits, pigs, cows, and other livestock. As the Squatter community matured, its ability to provide for itself increased. In some cases, Squatter families were relatively self-sufficient. Others augmented their gardening with odd jobs, income from the sale of art or other special skills, or by enrolling on welfare. It was not utopia, however. Few if any Squatters were overweight.

For the true hippie, however, the Squatters evolved into something resembling the original vision of the Cloud Farmers:

> It was definitely a special place in time. It was an outpost, a settlement in the middle of a wilderness, only the wilderness was created by the Corps of Engineers. It was secluded enough so that the people who were there could take the time to get into their own direction, their own self-sufficiency and to also form their own beliefs and communal ideas. There was a certain flow that went through the valley and it was almost like being connected to another place, another time, or another reality. It was a real movement—alternative from the word go.[6]
>
> It was a chance of a lifetime: to live on a beautiful river with no neighbors, miles and miles of land, and anybody around was just friends. You could try out all your dreams of the time.[7]

Life as a Squatter may have had its dreamlike qualitites, but it had its nightmare side, too, particularly to the outside world. Most local residents of the conservative, rural area had heard of hippies from the news media. Suddenly, down at the local supermarket or walking through the valley were dozens of long-haired men with bandanas, women with long paisley skirts, and racially mixed couples straight out of the pages of *Newsweek*. At best, local residents resented the Squatters for taking welfare and living in houses taken by the federal government from their former owners. At worst, there was a genuine hatred of the alternative life-style and its practitioners.

It got worse. During the summer, the summer youths (called Avenue B'ers by the Squatters) joined the Squatters. Crime, particularly petty theft and arson, increased. Rightly or wrongly, this was blamed on the Squatters. Drug use was also on the increase (as it was everywhere), and this was also blamed on the Squatters. In the valley itself, various acts of innocent but outrageous behavior were perpetrated on passing motorists, and confrontations occurred between the hippies and hunters, fishermen, law enforcement officials, and others. Some local residents living near the Squatters began fearing them, an additional burden in an area already suffering from the chaos of the land-acquisition process. Many local residents began traveling in groups and with weapons.

The Squatters had much to fear from the local residents, too. Local "red-necks" took to riding through Squatter territory and shooting at pets and houses.[8] At least two Squatters were wounded by unknown snipers, and in 1971 the Squatters petitioned Monroe County for police protection because of the harassment from local rowdies. In the fall of the same year, two Squatter barns filled with winter provisions were burned by unknown arsonists. Many Squat-ters began carrying arms and taking other measures to protect themselves, even though to do so violated the hippie's nonviolent philosophy.

Three federal agencies had to deal with the Squatters: the Corps of Engi-neers; the Justice Department, with its attorneys and federal marshals; and, to a lesser extent, the National Park Service. The Commonwealth of Pennsylva-nia was noticeably absent from the Squatter situation. The federal govern-ment had wanted the state to exercise jurisdiction in the matter since it could evict trespassers much more quickly than could the federal government.[9] This method had worked quite effectively in New Jersey, but the Pennsylva-nia Justice Department decided to stay out of the affair.

In 1971 the federal government began moving against the hippies living in Pennsylvania. Leases were allowed to expire, and in August trespass notices and court summonses were served. The Squatters, however, pooled their money and hired an attorney. The attorney obtained a continuance of the eviction cases until September 17, 1971.

While the cases were continued in one court, the Corps got a second federal judge to order federal marshals to "remove river people or squatters from three tracts of land consisting of 263.56 acres of land in Monroe and Pike Counties."[10] In the early morning hours of September 4, federal mar-shals raided the Squatter community. The surprise raid routed Squatters living in six houses, and these plus two barns and a tent campsite were immediately demolished by waiting bulldozers. The noise woke up the other Squatters. They climbed with their children up on rooftops and stood in the path of the bulldozers to stop further demolition.

Following the surprise raid, the Squatter Parents Association announced that the Squatters would leave the valley when Congress and President Nixon approved the construction of Tocks Island Dam.[11] The statement reflected an emerging belief in the Squatter community that resistance to eviction was also a stand against dam construction. In their short stay in the valley, the Squatters had come to love the beauty and serenity of the place, and they, like the local residents who had preceded them, did not want to see the valley drowned by a reservoir.

Following the 1971 raid, the federal government served various Squatters with criminal-trespass notices. Each demanded a jury trial. The Corps also got the public utilities to cut off service to Squatter homes so that they had no heat or hot water during the cold Minisink winters. These steps and others strengthened the Squatter resolve and bound the community closer.

In November 1972 the Justice Department filed suit in the U.S. District Court to evict 171 named Squatters. Many of the 171 persons were summer hippies who were long gone from the valley. The suit sought a court order requiring the Squatters to vacate the federal land and a permanent injunction to prevent them from returning. On November 16 armed federal marshals, state troopers, county sheriffs, National Park Service rangers, and Corps officials descended on the valley to serve court-appearance notices to as many Squatters as they could find. The process serving began peacefully but turned ugly. A scuffle broke out when a male Squatter tried to prevent a federal marshal from forcing his way into a female Squatter's house. Taunted and spat upon, the marshals called in state police equipped for riot control to serve the remaining notices.

The notices gave the Squatters twenty days to either leave or request a hearing. Since the Squatters could no longer afford an attorney, the Squatter Parents Association appointed an eight-person committee to represent them. In December 1972 the Squatters petitioned the court for:

1. An explanation of the Corps' land-acquisition process and condemnation proceedings in the valley;
2. A legal description of the property involved (the Squatters' petition alluded to technical violations in the Corps' original property purchasing);
3. and 4. Specific details about how they had interfered and were interfering with the administration of the Tocks Island Dam Project; and,
5. An explanation of the acts of irreparable damages they had caused (these had been alluded to in the federal suit).

As everyone anticipated, the case began to drag through the federal court system. Various legal maneuvers were tried by both sides, and the Squatters managed to hang on during the summer and fall of 1973. At each step in the legal process, however, they came closer to eviction.

The Squatter legal team brought many Squatters with them to Wilkes-Barre. Their appearance became a media happening, with long-haired men, barefoot women with suckling babies, dogs, and all the other trappings of hippie life. National and regional news media began taking notice of the Squatters as a part of a larger Tocks Island problem.

Both the Squatters and the federal government feared that the situation would end with violence. Through 1973 federal and state officials became increasingly worried that the Squatter community would physically resist eviction when the federal marshals finally moved into the valley. Adding to their fears was the armed defense of federal land by Russell Means and his American Indian Movement at Wounded Knee, South Dakota. The military minds of the Corps easily imagined a similar confrontation with the Tocks Island Squatters. In June 1973 the Corps hosted a closed meeting in Washington to brief the local congressional delegation on the seriousness of the situation. [12]

Living the squatter life in the valley had produced its own paranoia. The Squatters did not trust the law enforcement officials and perceived them to be filled with hatred. They expected the evictions to be brutal, and some expected to die. By the end of 1973, tensions had increased dramatically on both sides. The free-spirited life of a Squatter had become deadly serious. Many were ready to leave the valley, except for their commitment to the community and their opposition to Tocks Island Dam. [13]

The remaining months of the Squatters saw a gradual deterioration of their legal position. On November 12, 1973, Judge Michael H. Sheridan ruled that the Squatters were indeed illegally occupying federal property, and a summary judgment was issued giving the Squatters thirty days to leave. The Squatters indicated that they would appeal the decision. By mid-January 1974, however, the papers allowing federal marshals to carry out the evictions were being drawn up.

The Squatters filed their appeal with the Third Circuit of the U.S. Court of Appeals in Philadelphia. When this occurred, Judge Sheridan stopped the eviction process. The Squatters, believing they had won some more time, rejoiced. Charles Evans, one of the Squatter spokesmen, was even more optimistic, exclaiming, "We're going to win this case yet." [14] The Squatter legal team then took some time off before the next step in the legal process.

Unknown to the Squatters, Sheridan had signed the papers giving the federal marshals permission to evict them. Ninety marshals were brought into the area from the Midwest, and a backup force of dozens of Pennsylvania state police was readied. In paramilitary fashion, the marshals swept into the sleeping Squatter community on the bitterly cold morning of February 24, 1974. The marshals were armed with tear gas and pistols and were wearing bulletproof vests. The raid was not a complete surprise to the Squatters. Barry Kohn, a Pennsylvania state attorney, had driven from Philadelphia during the night to warn the Squatters of the attack. Kohn was afraid that a surprise attack would spark an armed confrontation.

The eviction process was highly organized and efficient. State police first sealed off the area to prevent access by the news media, an indication of the seriousness of the paramilitary action. The marshals then entered the area. Potential troublemakers had been identified beforehand, and these were rounded up quickly and handcuffed. Eleven Squatters were forcibly detained, but most were treated with courtesy. In less than one hour, it was all over. After some cursory attempt was made to remove personal belongings, federal bulldozers leveled all the buildings. Dozens of Squatter animals were left to wander in the area.

The Squatters were deposited outside the federal boundaries and prevented from returning. An immediate problem faced by the shocked Squatters was the need for shelter from the bitter cold. One of the women had delivered a baby during the previous night, four other women were pregnant, and there were babies and young children in the group. Eventually, the Squatters drifted away from the valley and found shelter and other assistance. Many finally wound up at Sweetwater Farm near Marshalls Creek, which became a refugee camp of sorts. There was brave talk of taking some kind of guerrilla action, or of suing the government for the loss of personal possessions and the "illegal" evictions. The days of the Squatters were over, however.

The Squatter eviction generated a great deal of bad publicity for the Corps of Engineers. Many persons wanted to believe that the evictions had been illegal and unnecessarily brutal since the Squatters' case was still being appealed. Editorials to this effect were published in many newspapers. According to Pennsylvania officials, however, there were good reasons for the quick eviction. They believed that local vigilantes were planning to rout the Squatters.[15] If this was so, the evictions may have avoided bloodshed in the peaceful Minisink Valley.

As a direct consequence of the Squatters, the Corps accelerated the razing of buildings in the Minisink Valley. In its zeal to prevent the Squatters from returning, the agency made some big mistakes. One house was destroyed by accident, historic buildings were demolished, and the Zion Lutheran Church was stripped of its floors, windows, and woodwork. After the church was gutted, Nancy Shukaitis, the Delaware Valley Conservation Association, the Sierra Club, and the New Jersey Public Interest Research Group obtained a restraining order stopping all further demolitions. The wanton destruction of historic buildings generated even worse publicity than the evictions.

The problems with the land-acquisition process and the Squatters were minor parts of the Tocks Island Dam controversy. Both, however, attracted a great deal of news coverage at a time when the dam was receiving other adverse publicity. The anguish in the Minisink demonstrated major problems with the Corps' operations in the valley. If the federal government had trouble managing its land-acquisition process and a band of "flower children," could it be trusted to deal with eutrophication or any other environmental problem?

13

Decisions Are Made

The Squatters were yesterday's business by the time the congressionally mandated Tocks Island review study got under way. The study, formally entitled *The Comprehensive Review Study of the Tocks Island Lake Project and Alternatives*, began in earnest in September 1974. Nobody was particularly thrilled with the prospect of another study. The Corps, the Delaware River Basin Commission staff, and other Tocks proponents thought that the study was unnecessary. The environmentalists, on the other hand, distrusted the Corps and doubted that any study could prove that Tocks Island Dam had merit.

During the fall of 1974, the Corps began developing a plan of study and screening potential consultant firms. While the study was being formulated, three information meetings were held to get public input. Both New Jersey and New York State appointed citizen advisory committees to watch the study and advise the governors on the dam decision.

While the review study was capturing all the headlines, the long-awaited supplement to the Corps' environmental impact statement was released. The 450-page document was far more comprehensive than anything that had preceded it. Particular pains had been taken to address the issues raised by the Council on Environmental Quality and the State of New Jersey. The report, however, was too late to influence any decisions on the dam.

In December 1974 the consultants for the review study were selected. URS/Madigan-Praeger, Inc., an engineering firm, and Conklin and Rossant, an architectural firm, were awarded a $1.15 million contract for what was now a six-month study of the Tocks Island Dam Project. Eventually, nearly a dozen subcontracts were awarded by the consultants to other firms and individuals.

The study was organized into five parts. Collectively these covered all the issues surrounding the dam.

Part A: *Analysis of Service Areas and Resource Needs* was to assess the demands for the four Tocks Island Dam purposes (water supply, recreation, flood control, and power) under three alternative growth projections.

Part B: *Review of the Tocks Island Lake Project* was a technical and historical examination of the engineering and planning that had been performed by the Corps, the Delaware River Basin Commission, and others involved with the design of the project. Included in Part B were economic analyses, water-quality impacts (such as eutrophication), major environmental impacts, and a review of the concerns raised by the dam opponents.

Part C: *Analysis of Alternatives to Supply Resource Needs* was to evaluate alternatives to the dam.

Part D: *Institutional Alternatives* was to evaluate the constraints imposed by the 1954 Supreme Court Decree and the Delaware River Basin Compact on the potential deauthorization of the dam project.

Part E: *Land Use and Secondary Effects of the Tocks Island Lake Project* was to evaluate the secondary impacts, such as growth and development, highway needs, social and life-style issues, and others.

The extensive public information program mandated by Congress was the responsibility of the Delaware River Basin Commission. As part of its assignment, DRBC set up regional depositories to ensure that the public had access to the same information being reviewed by the consultants. Even the most ardent Tocks watcher was impressed by the sheer volume of this material.

As the review study headed into 1975, one notable thing occurred. The Ford administration budget proposal for fiscal year 1976 contained nothing for Tocks Island Dam. It was the first budget proposal since 1964 to contain no Tocks money. The zero budget request reflected the Corps' philosophy that no more money was to be spent on the project until dam construction was assured.

In the spring of 1975, drafts of the five review study parts were released. After each part was released, the Delaware River Basin Commission held public information meetings. Each meeting was a one- or two-day marathon dominated by critical and scathing testimony from the environmentalists. The Save the Delaware Coalition, the Delaware Valley Conservation Association, the various Sierra Club chapters, and many others took turns finding errors, calling for more information, questioning the impartiality of the study, or generally blasting the study and the dam project. It was clear from the meetings that the dam opponents could not be appeased.

One of the most formidable opponents during the public meeting process was Mina Haefele, the head of the Delaware Valley Conservation Association. During the course of the study, she had spent hundreds of hours studying the project and had visited the New York City offices of the consultants. There she was given access to the consultant files and had developed contacts among the study employees.[1] As a result, Haefele generally knew many weaknesses in the study. These she used with relish at the public meetings and wherever else she went.

The public meeting process brought out a few original suggestions, too. Perhaps the most original suggestion came from Robert M. Brooks, a Lenni Lenape Indian. Brooks made an intelligent argument for giving part of the recreation area back to the Indians so that they could reestablish their former way of life. A runner-up for originality was the gentleman who claimed to represent gold mining claims on Kittatinny Mountain.

During the early months of the study, one new issue had emerged. The Medical Society of New Jersey had come out in opposition to the dam on the premise that the upstream poultry and cow wastes would introduce high levels of *Salmonella* and other pathogens into the reservoir. This issue was traceable to an internal U.S. Environmental Protection Agency memorandum that had been written in 1971.

Following the public meetings, the review study report was finalized. By early July the United Parcel Service was delivering copies of the final report to the doorsteps of dam opponents and supporters throughout the Delaware River Basin states. More than three-thousand pages of Tocks Island Dam material weighing eighteen pounds awaited the eager Tocks Island reader. The report volumes were attractively bound in gray and black, with brilliant orange chapter separators. In the haste to get the reports out, some pages had been printed upside down or collated in reverse, and, with use, the weak binding had a tendency to eject pages. Still, the consultants' study was the last and most comprehensive report concerning the decades-old main-stem dam proposal.

Anyone who thought that the review study would answer the many issues swirling around Tocks Island Dam was both wrong and naive. The study had promised to be impartial and objective. As a result, it was ambiguous, possibly reflecting the ambiguity of the dam project itself. The consultants found that Tocks Island Reservoir would become eutrophic *but* that this would not affect any reservoir purposes except some of the recreational uses. Growth impacts in the region surrounding Tocks Island would occur without Tocks Island Dam, *but* the townships nearest the reservoir would suffer inordinate burdens and require assistance. There were alternatives to Tocks Island Dam, *but* all of these had political, institutional, and environmental questions that required resolution. Tocks Island Dam was found to be more cost-effective than any combination of alternatives, *but* the concentration of all the purposes into one single project created some highly adverse impacts on the surrounding region and the environment. So it went.

In developing the review study, the consultants had relied on readily available information and data. Time did not allow much more than this. Nevertheless, the study was not a mere regurgitation of past work. In some areas, such as eutrophication, water supply, and estuary salinity control, new analytical techniques were used. The study certainly had weaknesses, too, but at a minimum it presented for the first time in one document all the issues, alternatives, impacts, projections, rationales, and ramifications of the Tocks Island Dam Project.

On July 22 and 23 the Delaware River Basin Commission held the final public meeting in the study process. At this time, DRBC was seeking public input for the governors' decision on the dam. Congress wanted DRBC to make its decision by the end of summer, and a summit meeting of the governors was already scheduled for July 31. At this meeting, the governors were either going to discuss how the decision was to be made or actually make the decision. The two-day public meeting was seen as the last chance to make comments and vent feelings before the final decision was made. During the two-day hearing, almost one-hundred people and organizations offered widely divergent opinions on the dam, and another twenty-nine submitted testimony for the record.

What was said at the public meeting mattered little. For the most part, everything had already been said during the past five years. The environmentalists were still solidly against the dam. The business, labor, engineering, and water interests were clearly for it. The $1.5 million study had changed no minds. In any event, the voluminous testimony would have little impact on the decision of the Delaware River Basin governors. This decision would come quicker than most people anticipated.

On July 31 the Delaware River Basin Commission gathered in Newark, New Jersey, for its summit meeting. Three governors attended the meeting: Shapp of Pennsylvania, Byrne of New Jersey, and Tribbitt of Delaware. Joining them were Ogden R. Reid, for Governor Carey of New York, and Thomas F. Schweigert, for the federal government. During the morning, the commissioners debated, sometimes heatedly, about what to do about Tocks Island Dam. It became clear that a decision could be reached. When the vote was taken in the afternoon, construction of Tocks Island Dam was defeated by a 3 to 1 vote.

The sole state voting for the construction of Tocks Island Dam was Pennsylvania. Governor Shapp and his chief advisor, Maurice Goddard, were firmly convinced that the dam was vitally needed for flood control and water supply. Goddard was the state's Secretary of Environmental Resources and had headed the Department of Forests and Waters back when that agency was planning to build a state dam at Wallpack Bend. He was one of the few basin water experts who had been around for the 1954 Supreme Court decision, the disastrous 1955 flood, and the record drought of the mid-1960s. As a result, Goddard was an unswaying supporter of Tocks Island Dam.

Voting against construction were New York, New Jersey, and Delaware. New York's position had been known for some time. It did not want to pay for any pollution abatement measures that might be required as a result of the construction of a downstream reservoir. Delaware, on the other hand, had only a passing interest in the dam. Its governor had indicated in the past that he would vote with the majority. The federal commissioner abstained from voting on the basis that the federal government was the project sponsor. His vote did not matter anyway since Congress wanted to hear from the states.

New Jersey's position had been uncertain ever since Governor Byrne had taken office. Eutrophication and the newer *Salmonella* issue concerned the state, but it also had definite water-supply needs. The data in the review study report, however, suggested that New Jersey could get by without Delaware River water for at least twenty years if water supplies were developed in other parts of the state first. Although four states voted on the construction of Tocks Island Dam, it was New Jersey's vote that decided the issue.

After a half century in the making and a decade of controversy, was Tocks Island Dam now dead? In the words of a bitter Governor Shapp, Tocks Island Dam was "mortally wounded and in danger of dying."[2] It was not dead, however. On the books at least, Tocks Island Dam was still a congressionally authorized dam project and a key component of the DRBC Comprehensive Plan. Moreover, two of the Delaware River Basin states were still interested in the dam—Pennsylvania immediately and New Jersey within two decades. Most of the land for the dam had been acquired, and the design of the dam lay on the shelves of the Corps of Engineers. Tocks Island was indeed a phoenix waiting to rise out of the ashes of 1975.

Deauthorization of Tocks Island Dam was the next item on the environmental agenda. Bills to deauthorize the dam project had first been introduced in Congress in 1974, but they had gotten nowhere. With DRBC voting against construction of the dam, the door seemed open to try again.

The consultants for the review study had examined three types of issues inherent in the deauthorization of the Tocks Island Dam Project.[3] First, what would happen to the land acquired by the Corps of Engineers? By mid-1975 the Corps had spent approximately $50 million for seventeen-thousand acres in the Minisink Valley. Unless Congress decided otherwise, this land could be disposed of as surplus federal property since it was no longer needed for its original purpose.

The second issue was the legality of the Delaware Water Gap National Recreation Area and, to a lesser extent, the Kittatinny Mountain Pumped-Storage Project if Tocks Island Dam was deauthorized. Both were linked by federal legislation to the dam project. Did its deauthorization automatically deauthorize the recreation area? If the DWGNRA was deauthorized, what would happen to the 31,043 acres of parklands that had been acquired by the National Park Service? And what good were the DWGNRA's 31,043 acres if the National Park Service couldn't get ownership of the Corps' land? The National Park Service land was above the intended reservoir line and not near the river.

The final issue was the impact of deauthorization on the provisions of the 1954 U.S. Supreme Court Decree. One of Pennsylvania's conditions for agreeing to the water diversions of New York City and New Jersey was the latter's support for the construction of a dam at Wallpack Bend. Tocks Island Dam had, of course, replaced the dam project mentioned in the decree—or had it? Was New Jersey's anti-Tocks vote the same as withdrawing its support for the Wallpack Bend dam mentioned in the Supreme Court Decree? If Tocks and Wallpack Bend were one and the same, was New Jersey's right to divert Delaware River water now invalidated?

All the questions surrounding the deauthorization of Tocks Island Dam were extremely serious, with long-range implications.

In the days preceding the Delaware River Basin Commission vote, Brig Gen. James L. Kelley, the head of the Corps' North Atlantic Division, had stated that if DRBC voted yes dam construction should begin immediately. If they voted no, Kelley planned to recommend that Tocks Island Dam be deauthorized. According to Kelley, "The project should not be deferred, but rather should be deauthorized so that alternative plans to meet the needs of the region may be developed."[4] The Corps feared that the existence of a deferred but still authorized project would be a psychological impediment to solving the region's water needs. In September 1975 the Corps of Engineers joined the Council on Environmental Quality, the U.S. Environmental Protection Agency, the Office of Management and Budget, the National Park Service, and other agencies in calling for the deauthorization of Tocks Island Dam.

Senate Bill S. 3106 was introduced on March 9, 1976, by Senator Clifford Case of New Jersey and six other senators from the Delaware River Basin states. Case's bill had three purposes: to deauthorize the dam, to transfer the Corps' land in the Minisink to the National Park Service, and to authorize the National Park Service to relocate U.S. Route 209 outside the park boundaries. The highway relocation had been part of the dam project because the reservoir would have flooded the highway between Bushkill and Milford, Pennsylvania, but it was thought necessary even without the dam because of increasing traffic and highway noise.

The Senate bill reached the public-hearing stage in July 1976. Veteran watchers of Congress, however, realized that the bill was moving too slowly to be enacted in that session of Congress. A major problem was the highway relocation part of the bill. Opposition to the highway relocation stemmed from the belief that the Corps would get the job. There was sentiment in the Senate for throwing the Corps completely out of the valley, even if it jeopardized the highway relocation project.

On July 23 and 26 hearings were held by the Senate Public Works Committee's Subcommittee on Water Resources. Lining up on both sides of the question for one more time were the various adversaries in the Tocks Island Dam controversy. There were changes, however. Joining the environmentalists in the call for deauthorization were the Corps of Engineers and the National Park Service. Absent from the debates was a long-time Tocks supporter, the Delaware River Basin Commission. Times had certainly changed.

Among the four Delaware River Basin states, only New York favored deauthorization. New Jersey and Delaware were against it, believing that deauthorization should wait until alternative water-supply and flood-control measures were developed. Pennsylvania was opposed to deauthorization and pushed for the immediate construction of the dam. Pennsylvania's representative, Maurice Goddard, did not believe that environmentally suitable alternatives could be found.

Among the opponents of Tocks deauthorization was Representative Frank Thompson of New Jersey. Thompson, the chairman of the House Administration Committee, vowed to use his considerable influence to kill the House version of the bill. He was successful. Soon afterward, the Senate bill was also killed.

In early 1977 deauthorization bills were introduced again in both the House and Senate. The U.S. Route 209 issue was resolved by leaving it out of the bills. The 1977 bills met the same fate as their predecessors when they died in the House and Senate Public Works committees. Neither committee was particularly sympathetic to environmental causes or the deauthorization of dam projects. After both bills died, it became clear that a method of circumventing the two committees was needed.

Back in 1968, Congress had passed the Wild and Scenic Rivers Act (P.L. 90-542), which established a National Wild and Scenic Rivers System that initially included eight river sections. Twenty-seven other river reaches were designated by the act as candidates for the system, pending study. Rivers selected for the system were to possess "outstanding remarkable scenic, recreational, geologic, fish and wildlife, historic, cultural or similar values."[5] Depending on the degree of development adjacent to them, the rivers could be designated as wild, scenic, or recreational.

The Wild and Scenic Rivers Act had been developed by the Johnson administration. In 1966 Johnson had called for a "National Wild Rivers System" that would "identify and preserve free-flowing stretches of our great scenic rivers before growth and development makes the beauty of the unspoiled waterway a memory."[6] The key phrase in the statement was "free-flowing," as the system was designed to complement (or counter) the proliferation of federal dams being authorized in the 1960s. The act did not preclude future dams on designated reaches but presented an impediment to their construction.

To the frustrated supporters of Tocks deauthorization, the wild and scenic rivers legislation looked like a way of obtaining de facto deauthorization. The dam project would remain authorized, but its construction in the future would be severely hampered. Even better, the scenic rivers legislation would bypass the House and Senate Public Works committees; thus, it had a chance for passage.

Part of the impetus for using the Wild and Scenic Rivers Act was that a portion of the Delaware River was soon to be included in the system. The 1968 legislation had included the Delaware River upstream from Port Jervis as one of the twenty-seven potential candidates for designation. By 1977 the studies leading to the designation of the Upper Delaware Scenic and Recreational River were nearing completion.

Including the "Middle Delaware" in the National Wild and Scenic Rivers System appeared to be a good idea to the anti-Tocks forces. The scenic qualities of the Delaware River in the DWGNRA supported its inclusion, and, moreover, the action would place a major roadblock to the future construction of Tocks Island Dam. It was not the same as deauthorization, but it could have almost the same effect.

In May 1977 freshman congressman Peter B. Kostmayer, from Bucks County, Pennsylvania, introduced legislation in the House that incorporated both the upper and middle sections of the nontidal Delaware River into the National Wild and Scenic Rivers System. On November 1, 1977, Senator Case introduced a companion bill in the Senate. A hearing on the Kostmayer bill was held by the House National Parks and Insular Affairs Committee in November. The hearing was held at New Hope, Pennsylvania, in Kostmayer's district and was chaired by Representative Phillip Burton (D-Cal.), the committee chairman. About 350 people attended the hearing, and many spoke in favor of the middle Delaware designation. Four persons, led by Mina

Haefele, bicycled to the hearing from the Tocks Island region to present a petition with three-thousand signatures supporting designation. Burton was impressed with the response, and he promised to support Kostmayer's bill. Burton was one of the most influential House members, particularly in national parks matters, and his support was critical.

Kostmayer's bill was subsequently incorporated into an omnibus parks and rivers bill (H.R. 12536) that was voted out of the House Interior and Insular Affairs Committee on May 15, 1978. The bill was entitled the National Parks and Recreation Act of 1978, or the "Park Barrel Bill" by its critics. It provided parks and recreation projects in forty-four states, three territories, and over 48 percent of the congressional districts in the nation.

House debate on the recreation package was completed in June. Shortly afterward, more than seventy amendments were offered to the bill. Most were technical changes. On July 10, however, Representative Frank Thompson offered an amendment to strike two sections from the bill. These sections would transfer the Corps' land in the Minisink to the National Park Service and include the middle Delaware in the National Wild and Scenic Rivers System. Thompson wanted these sections removed because they would impede the future construction of Tocks Island Dam.

The debate on the Thompson amendment was lengthy and lively. On one side were Thompson, a highly seasoned and influential congressman, and his followers. The creation of the Tocks Island Dam Project, the Delaware River Basin Compact, and the DWGNRA were at least partially the results of his efforts. In his mind, the dam was unfinished business, and he had supporters in Congress to back him. Many of these represented districts in the lower Delaware Valley that wanted Tocks water for their cities and industries.

Calling Kostmayer's proposal to designate the middle Delaware as a scenic river a "back door process," Thompson raised the specter of thousands being affected by floods and millions being without water during a drought. Thompson's supporters raised other arguments. Some felt that Tocks water would bring industrial prosperity back to the dying Northeast, or that it would somehow save the groundwater resources of New Jersey's unique Pine Barrens. Floods, droughts, state water rights, and the blatant circumvention of the Public Works Committee were all raised.

Leading the attack on the Thompson amendment was Peter Kostmayer, who referred to Thompson as one of his political heroes. The freshman congressman, however, was ready with facts and eagerly offered rebuttal to arguments concerning salinity, flood control, water supply, and the continuing search for Tocks alternatives. As promised, Representative Burton supported the middle Delaware designation and Kostmayer's arguments. Overstating somewhat, Burton declared that "this is an idea whose time has come" since it would provide recreation for "millions upon millions of people [who] are living literally stacked like cordwood."[7]

Kostmayer's arguments favoring designation did not hide his attempts to deauthorize Tocks Island Dam. When queried why his proposal did not go through the Public Works Committee, Kostmayer replied, "That is not relevant. Let us not have any parliamentary gymnastics. Let us resolve the issue and stop the dam and save the taxpayers of this country $1 billion. That is the issue." When asked again why he should deny the Public Works Committee the power to deauthorize the dam, Kostmayer replied sarcastically, "God forbid that we should deny the Public Works Committee anything."[8] Kostmayer's appeal was directed toward a coalition of newer congressmen, who were less bound by House tradition, and fiscal conservatives looking for money to save. For a freshman congressman, Kostmayer's actions were bold and well planned.

Joining Kostmayer and Burton in the move to strike down the Thompson amendment was a host of other supporters. Most had been actively involved in the past effort to get the dam deauthorized. One of Kostmayer's key supporters was Congresswoman Helen Meyner of New Jersey, the wife of former governor Robert Meyner. Ten years earlier, Robert Meyner had been damned by the environmentalists for selling Sunfish Pond to the power companies. Mrs. Meyner distributed maps and photographs of the Tocks Island region to each congressman and called on her colleagues to "save the last free-flowing river in the Northeast." According to Meyner, "The last thing we need is another hearing, another meeting, another report on Tocks Island Dam. There has been more ink spilled over the issue than would fill any lake created by the dam."[9]

After the lengthy debate, the House voted on the Thompson amendment. It was defeated by a vote of 275 to 110. On July 12 the final version of the recreation bill passed the House by a 341 to 61 vote. A major hurdle had been overcome.

In October 1978 the National Parks and Recreation Act came up for a vote in the Senate. At the last minute, Senator Mike Gravel of Alaska made a motion to exclude the middle Delaware's scenic-river designation. This surprise move was attributed to Frank Thompson, who had come over from the House to collect his many IOUs. The amendment was defeated in the Senate by a 3 to 2 margin, however, and the bill was passed. President Carter signed it into law on November 10, 1978. Both the upper Delaware and the middle Delaware were now components of the National Wild and Scenic Rivers System.

On December 16 the opponents of Tocks Island Dam held a victory celebration. A dinner was sponsored by the Save the Delaware Coalition, the Delaware Valley Conservation Association, and the Four County Task Force, a group formed during the demise of the Tocks Island Regional Advisory Council. It was a time for toasts and awards. The banquet's host, Harold Lockwood, Jr., received an award from Trout Unlimited for his involvement in the dam fight. Nancy Shukaitis was inducted into the Pennsylvania Fish Commission's "Order of the White Hat," while Peter Kostmayer received an award from the Save the Delaware Coalition. He was praised as "David," with the Corps depicted as "Goliath" in the Tocks battle, even though this analogy ignored the Corps' support for dam deauthorization. Among the special guests were two Lenni Lenape Indians whose ancestors had had their own problems in the Minisink.

The inclusion of the middle Delaware in the National Wild and Scenic Rivers System was definitely a victory for the environmentalists. They had every reason to celebrate. Little joy was found within the Delaware River Basin water community, however. The de facto deauthorization of Tocks Island Dam was threatening to send the Delaware River Basin states back to the U.S. Supreme Court for the third time in fifty years.

14

A Matter of Good Faith

The relationship of Tocks Island Dam to the 1954 Supreme Court Decree had first been raised by the Corps of Engineers in October 1974. At that time, the Corps indicated that dam deauthorization could lead to a reduction in New York City's water diversion. The attempt to link the dam project to the city's water rights was a transparent effort to change New York State's anti-Tocks position. New York City considered the Corps' statement "a bolt from the blue."[1] The city discounted the threat, believing that its diversion was protected by the Delaware River Basin Compact. The Corps quietly dropped the matter, but the issue was kept alive by Carmen Guarino, Philadelphia's water commissioner, and others.

New Jersey's diversion rights were more closely tied to a main-stem dam. The Amended Decree required the state to support a dam at Wallpack Bend. Were Tocks Island Dam and the Wallpack Bend Dam mentioned in the Supreme Court Decree the same dam? Finding out involved reopening the Delaware River Case. Not only would this demonstrate the states' failure to work together via the Delaware River Basin Compact, but it presented a Pandora's box of unknowns.

Two reports on both sides of the issue were prepared by private parties. In January 1976 the pro-Tocks Delaware Valley Council published a nineteen-page *Research Memorandum* on the matter. The study outlined the history of the Wallpack Bend and Tocks Island dams and concluded that they were one and the same. New Jersey's vote against dam construction was therefore seen as a violation of the Supreme Court Decree. Based on this, the council recommended that Pennsylvania ask New Jersey to either rescind its 1975 Tocks vote or cease diverting water from the Delaware River. If New Jersey did not comply, the council recommended that Pennsylvania seek redress in the Supreme Court.

The position of the Delaware Valley Council was rebutted by the Save the Delaware Coalition. The coalition's lawyers concluded that New Jersey had never officially substituted Tocks Island Dam for the one at Wallpack Bend; thus, the decree provisions did not apply. As the pro-Tockers were quick to point out, this narrow interpretation ignored the fact that Tocks Island Reservoir would have put Wallpack Bend under many feet of water.

DRBC's fifteen years of planning were undergoing review as the debates continued. In late 1976 the agency had received a $1.1 million grant (matched with $400,000) from the U.S. Water Resources Council to reevaluate its comprehensive plan (the so-called Level B study). Other planning was taking place, too. State water-supply plans were being prepared in Pennsylvania and New Jersey, while DRBC and the Corps had salinity and nonstructural flood-control studies under way. In one way or another, each of these studies was seeking alternatives to Tocks Island Dam. Until they were completed, however, no one could guarantee that alternatives to Tocks Island Dam existed.

The Delaware River Basin Commission and its staff had been faced with a dilemma during the debates on deauthorization. A position on deauthorization was not possible because of the differing opinions among the basin states. Adding to DRBC's dilemma was the question of Tocks Island Dam's relationship to the Wallpack Bend Dam of the Supreme Court Decree. Sections 3.3(a) and 3.5(a) of the Delaware River Basin Compact prevented DRBC from taking any actions affecting the decree unless there was unanimous consent among the decree parties. The Delaware River Basin Commission was therefore forced to be silent during the critical debates on deauthorization.

When the focus of the debates shifted to the scenic-river designation, the commission felt less constrained. At Senate and House hearings in 1977 and 1978, DRBC's executive director, Gerald M. Hansler, testified that DRBC wanted the middle Delaware's scenic-river designation kept separate from the upper Delaware legislation. DRBC felt that the middle Delaware should be evaluated on its own merits as a scenic river and that the evaluation needed to address the loss of Tocks Island Dam due to the designation. DRBC did not want the Tocks Island Dam site preempted before its planning studies were completed.

In June 1978 the Environmental Defense Fund released a report addressing the issues surrounding the middle Delaware's scenic-river designation. EDF pointed out that the Wild and Scenic Rivers Act contained a mechanism for building future federal dam projects on designated rivers. Thus, designation of the middle Delaware was not the same as deauthorization. From this EDF concluded that designation did not interfere with the 1954 Supreme Court Decree provisions.

The Shapp administration in Pennsylvania did not agree. When it appeared likely that the middle Delaware would be designated as a scenic river, the state began instituting legal actions. Prompting these actions was a letter from the Interior Department arguing that the federal government had the right to abrogate Pennsylvania's rights to the Delaware River. The Interior Department believed that designation of a scenic river in the Tocks region did, in fact, have the same effect as deauthorizing the dam.[2]

On June 30 Pennsylvania and the City of Philadelphia filed suit in the U.S. Eastern District Court against President Carter, Secretary of the Interior Cecil B. Andrus, and other Interior Department officials. The suit sought to stop the executive branch from implementing any legislation that substituted a scenic and recreational river for Tocks Island Dam. The plaintiffs contended that the designation had to be approved by the Delaware River Basin Commission in accordance with the provisions of the Delaware River Basin Compact, and that the designation short-circuited the provisions of the National Wild and Scenic Rivers and National Environmental Policy acts.

Before filing the suit, Governor Shapp officially asked Governor Carey of New York and Governor Byrne of New Jersey to support his state's position on the middle Delaware scenic-river designation. Shapp's letter to the other governors made Pennsylvania's intentions clear.

> I wish there to be no mistake as to Pennsylvania's position or intention in this matter. If the proposed legislation is enacted, thereby leaving Pennsylvania without the option of implementing an impoundment on the mainstem of the Delaware as contemplated by Paragraph V.A of the Decree, Pennsylvania will take immediate legal action to protect the Commonwealth's rights in the Delaware River.[3]

New York and New Jersey were not willing to support Pennsylvania's position. After the middle Delaware became a scenic river, Pennsylvania began pursuing the legal option expressed in Governor Shapp's letters. Pennsylvania believed that the chain of events surrounding the scenic-river designation made a good case for getting the Supreme Court to reduce both New York's and New Jersey's allowable out-of-basin water diversions.

The Commonwealth of Pennsylvania would not go to the Supreme Court, however. Governor Shapp's term in office was almost over, and both candidates for his job were opposed to Tocks Island Dam. In all likelihood, the next Pennsylvania governor would quietly drop the federal suit and not aggressively pursue the Supreme Court option either. Still, the state had concerns that needed resolving. It was time for a different tactic.

At the October 25, 1978, meeting of the Delaware River Basin Commission, Maurice Goddard presented a draft resolution to his fellow commissioners. The resolution was premised on Pennsylvania's contention that the middle Delaware's scenic-river designation "appears to impair substantially the arrangements for the equitable apportionment of Basin waters set forth in the 1954 Decree."[4] According to Goddard, Pennsylvania needed 300 to 600 million gallons of water per day. This water was not available without building Tocks Island Dam.

Goddard's draft resolution called for the Delaware River Basin Commission to invite the parties to the Supreme Court Decree to enter into "good faith negotiations." The purpose of the negotiations was to work out a new formula for the equitable apportionment of the basin's water. The formula for equitable apportionment would then be adopted by DRBC and submitted to the Supreme Court for approval. If the parties to the decree did not agree to the negotiations, or if the new formula was not passed unanimously, Pennsylvania planned to go to the Supreme Court.

In the next month and a half, Goddard's draft resolution was worked over extensively by all the involved parties; it was finally adopted on December 13, 1978. The language of the resolution differed substantially from Goddard's original draft, but the key ingredients were still there. DRBC was to invite all of the parties to the 1954 decree to "enter into serious good faith discussions to establish the arrangements, procedures, and criteria for management of the waters of the Delaware River Basin."[5] By March 1979 the talks were under way.

The parties to the Good Faith negotiations were different from the members of the Delaware River Basin Commission. One of DRBC's members, the federal government, was not a party to the decree, and New York City, a decree party, was not a signatory to the Delaware River Basin Compact. The four Delaware River Basin states were parties to both the decree and the compact, however, and most Good Faith meetings were held in conjunction with DRBC meetings.

The Good Faith negotiations were a private affair without public participation of any kind. However, DRBC's Level B study, funded by the U.S. Water Resources Council under the Federal Water Resources Planning Act of 1965 (P.L. 89-80), included extensive opportunities for public involvement. The study was nearing completion as the Good Faith talks began. Public workshops were held throughout the study, and a study advisory committee was established. Serving on the committee were representatives of many citizen organizations and government agencies. In fact, the committee list read like a virtual Who's Who from both sides of the Tocks Island Dam controversy.

After the Good Faith talks got under way, they were linked, to some degree, with the DRBC Level B study. Each activity fed information to the other, and the Level B study served as the forum for public input. The Level B study was also more comprehensive than the Good Faith talks since it dealt with other water-resource subjects as well. Water conservation, water supply, and flow maintenance were three of the elements of the Level B study that tied it to the Good Faith talks.

By the time that the Level B study got under way, environmental opposition had killed another Corps dam project in the Delaware River Basin— Trexler Dam on Jordan Creek in the Lehigh River watershed. Only 5 percent the size of Tocks, the project had been moving along behind the Tocks Island Dam Project ever since both had been authorized in 1962. In 1976 the project received funding for the start of construction.

In many respects, the Trexler Dam Project was victimized by Tocks Island Dam. First of all, the Tocks Island fight made the citizens of the region much more aware of the environmental impacts of dams in general. More important, however, was the fact that power companies that had been hoping to get water from Tocks Island Dam were now planning to use Trexler water instead. In the wake of the Tocks Island defeat, DRBC had directed the power companies to build water-storage facilities.[6] The pending construction of Trexler Dam, with no immediate customers for its water, led to a proposal that Trexler water be used by the power companies on an interim basis. The Lehigh Valley voters did not like this and voted three to one against dam construction. After the vote, state officials killed the project. The proposal appeared to many to verify the long-standing claim that public dams in the basin were being built exclusively for the private power companies.

DRBC had required power companies to build water-storage reservoirs so that water was available to offset the tremendous quantities of water consumed during the generation of electricity. The reservoirs would release water during the drier months when flows in the Delaware River fell. The projected increase in water consumption concerned the commission because water flowing into the tidal portions of the Delaware retards the intrusion of sea salts coming up the Delaware Estuary. Although it was by no means a certainty, water planners had assumed for some time that a flow of about 3,000 cfs was needed at Trenton to keep salinity levels away from Philadelphia's intake at Torresdale. The loss of Tocks Island Dam meant that its huge volumes of water would not be available during critical dry seasons.

By the mid-1970s salinity control had become the driving force behind most of the water-resources planning in the Delaware River Basin. The issue was old, yet it was new. The "problem" of salinity intrusion in the Delaware Estuary had first been raised during the 1929–31 Supreme Court debates,[7] which had taken place during a severe four-year drought. After the drought, both the Corps and Incodel examined the salinity issue. The Corps' 308 Plan had suggested the possibility of using reservoirs for salinity-control purposes, as did its Tocks Island Dam pool-level study thirty years later. In both cases the Corps decided that storing water for flushing salinity downriver was impractical. Incodel came up with similar conclusions during its studies in the early 1940s.[8] It had, however, included low-flow augmentation as part of its integrated water project of the 1950s.

The drought of 1961–66 dramatically changed the thinking concerning salinity. The drought was more severe and lasted much longer than the drought of 1929–32. The safe-yield assumptions of the various Corps dam projects and the New York City reservoir system had been based on the less severe drought. These assumptions were incorrect. After the 1960s drought, for example, it was determined that New York City could not meet its Montague flow releases and still take its allowable diversion of 800 mgd when a drought like that of the 1960s occurred. The safe yield of the system had decreased 40 percent.

The second factor making salinity a primary policy consideration after the 1960s drought was the fact that ocean-derived saltwater had seemingly threatened Philadelphia's water supplies. During the drought, higher than desired chloride levels had intruded within eight miles of the city's Torresdale intake. Chloride concentrations at the intake had also risen, although these probably emanated from local sources.

Another concern highlighted by the 1960s drought was groundwater contamination in South Jersey. The Camden metropolitan area relies on the Potomac-Raritan-Magothy (P-R-M) Aquifer for its water. Because of the extensive pumping of the aquifer, the P-R-M no longer flows toward the Delaware River. Instead, the groundwater flow has reversed, and the river can now enter the aquifer (i.e., the river recharges it). Although aquifers are less sensitive to salinity intrusion than an intake pipe, high salt levels can persist for a long time in the groundwater once the salt is there.

Concern over aquifer contamination was first raised in the 1940s, when the Delaware Estuary was extremely polluted. The first studies, begun in 1944, indicated that there was a possibility that the P-R-M could be contaminated by the pollutants and salts in the river. As a result, Incodel sponsored a conference in 1949 that led to a series of groundwater studies in the 1950s by the U.S. Geological Survey. By the end of the decade, these studies had verified that the Delaware Estuary recharged the P-R-M Aquifer under certain hydrological conditions. This meant that if ocean-derived salts were present in the river the pumps in the Camden well fields could pull them into the aquifer.

As a water-resource problem, salinity in the Delaware River Basin is very confusing. When pre-1966 Delaware River dam proposals referred to water-supply benefits, they were generally referring to direct benefits that would be obtained by supplying raw water to customers. The primary water-supply purpose of Tocks Island and the rest of the Corps reservoirs was to add water to the Delaware River system to equal the projected increase in water use.[9] After the 1960s drought, however, water supply came to mean something else entirely: the protection of intakes and groundwater aquifers from salinity intrusion. Philadelphia was not going to get an upland water supply from Tocks Island Dam, but rather drought protection. This was an important but extremely subtle change from the pre-1960s planning. Since drought protection was couched in water-supply terms (i.e., water supply, water-quality control, and low-flow augmentation), the shift in emphasis was not widely recognized.

During the debates on Tocks Island Dam, salinity control was made an emotional issue by the pro-Tockers. The specter of seawater entering a municipal drinking-water supply sounds alarming. In actuality, however, the seawater is extremely diluted and may not even impart a taste to the water. The water is not necessarily unfit for drinking (except for persons with serious health problems), and, at worst, the water is still available for fire protection and sanitation. It is not the same as being without water.

Some supporters of Tocks Island Dam also chose to ignore the fact that critical chloride levels never did reach the Philadelphia intake during the 1960s drought. This drought had a return interval estimated to range from several hundred years or more in the upper Delaware River Basin to one hundred years at the mouth of the Schuylkill River.[10] Thus, water stored for a record drought may rarely be used, if ever.

The drought of the 1960s stimulated numerous studies concerning the freshwater flows needed at Trenton to retard salinity intrusion in the Delaware Estuary. The last study conducted before the 1975 Tocks Island Dam decision was done by Dr. Charles Kohlhaas, an employee of URS/Madigan-Praeger. Using an advanced statistical model, Kohlhaas came up with a startling prediction: Salinity standards would be violated only once every one hundred years at Torresdale and only once every five hundred years at the mouth of the Schuylkill River. And that was the conservative case. Under less conservative assumptions, salinity standards were never violated. Even more startling, flow releases from Tocks Island Dam appeared to have little, if any, impact on salinity levels in the Delaware Estuary.

The Kohlhaas analysis severely damaged the case for Tocks Island Dam. The supporters of the dam and various water experts by no means accepted his conclusions. As a result, the Delaware River Basin Commission sponsored a seminar on the Kohlhaas work and invited more than two dozen experts in the fields of tidal hydrology, modeling, and Delaware River Basin management. The seminar was held too late to affect the Tocks Island Dam decision, but it did set the stage for future work on the salinity question. This work would form the basis for recommendations coming out of the DRBC Level B study and the Good Faith talks.

The primary conclusion of the seminar participants was that a deterministic, time-varying salinity model of the Delaware River system was needed. A computerized, mathematical model was thought to be the answer since it would be a better predictive tool for water-resources planning. Several Delaware Estuary models had been developed in the past, and the state of the art had advanced since then.

In the post-Tocks period, the Delaware River Basin Commission assigned the highest priority to resolving the salinity questions that had been raised by the Kohlhaas work.[11] Several things happened as a result. In 1976 DRBC hired Drs. M. L. Thatcher and D. R. F. Harleman to update and improve a Delaware Estuary salinity model (the Transient Salinity Model, or MIT-TSI model) they had developed while at the Massachusetts Institute of Technology. The model was completed in 1978 and, after testing, was used extensively by DRBC to evaluate alternatives for both the Level B and Good Faith activities.

DRBC also sought and obtained authorization from the House Committee on Public Works and Transportation for a Corps salinity study, which was initiated in late 1977.[12] The study eventually explored several different salinity issues and led to additional modifications to the MIT-TSI model. At the same time, a separate study by the Corps led to the development of a daily-flow simulation model of the Delaware River Basin. This model could predict the volumes of freshwater coming down the Delaware River under various combinations of existing and proposed reservoirs. The various models were valuable water-resource planning tools.

The new salinity model was used extensively by the Level B planning team. One of its tasks was to compare the dam projects in the DRBC Comprehensive Plan with six alternative projects developed by the Tocks Island review study and nine potential high-flow skimming projects. The purpose of the evaluation was to determine the cheapest combination of dam projects for meeting eleven potential flow objectives and their corresponding salinity levels. The alternative flows ranged up to a flow of 1,830 cfs above the basin's existing flow-maintenance ability. As the flow objective increased, more and more dam projects were needed. Up to eight dams were needed by the tenth alternative.

Only Tocks Island Dam and one other dam were needed to meet the eleventh, or highest, alternative flow objective. Because of its large size, Tocks Island Dam also had the cheapest water of any of the alternatives. If a lesser flow objective was chosen, however, Tocks Island Dam was no longer cost-effective, because it provided more water than needed. As a result, Tocks Island Dam was not recommended by the Level B planners. They believed that a less than maximum flow objective would suffice if water conservation measures were instituted.

The Level B planning group wound up its activities in October 1979 with the publication of a final report.[13] The plan recommended a variety of potential policies for reservoir construction, flow maintenance, salinity control, water conservation, water quality, and other water-resource subjects. The "nuts and bolts" of the so-called Preferred Plan called for a flow objective at Trenton of 2,840 cfs, plus a 160 cfs reduction in water consumption by downstream water users. This added up to 3,000 cfs, the historical flow objective at Trenton.

New water-storage facilities were recommended. Two were enlargements of existing flood-control dams: F. E. Walter in the Lehigh watershed and Prompton in the Lackawaxen watershed. The enlargement of these reservoirs had been part of the Corps' 1962 reservoir plan. A new project was also recommended. This was the Merrill Creek Project near Phillipsburg, New Jersey, a high-flow skimming reservoir being planned by the Delaware River Basin electric utility companies to make up water lost during power generation.

The Level B plan also recommended two other dam projects for possible construction. These were Hackettstown Reservoir in New Jersey and the enlargement of Cannonsville Reservoir in New York State. The construction of Hackettstown had been contemplated by New Jersey for many years, but the dam project was not being actively pursued in 1979. The enlargement of New York City's Cannonsville Reservoir had also been suggested in the past, but it was not being pursued in 1979 either.

Finally, the Level B study recommended that Tocks Island Dam be retained in the Delaware River Basin Commission's Comprehensive Plan and that some time after the year 2000 DRBC should decide once and for all whether the project should be built. This recommendation was not as strong as the environmentalists would have liked, but it was better than having Tocks Island Dam recommended for immediate construction.

The Level B plan was far from being dam-oriented. Most of its proposed policies were widely favored by the public. Particularly pleasing to the environmentalists were policies promoting nonstructural flood-control measures and water conservation. Water conservation was considered the cornerstone of the Level B plan, and water savings up to 15 percent by the year 2000 were strongly pushed by the plan. With these policies and others, the plan sought to prevent, or at least delay, the construction of a dam on the Delaware River.

In early 1977 a new issue had arrived at DRBC's door. New York City's reservoir operations were causing problems for the cold-water fish in the upper Delaware River. New York State wanted the problem corrected, and in May 1977 DRBC approved an experimental reservoir operation plan that hoped to correct the problem in the upper basin. DRBC's approval also called for the creation of a task force to develop drought management criteria for the New York City reservoirs.[14] During the onset of a drought, DRBC wanted flexibility that could maximize the water storage in the New York City reservoirs. By maximizing the available water, DRBC was hoping not only to increase the city's ability to cope with a drought but also to prevent a situation from arising in which the city could not meet its obligations to the downstream states.

Serving on the Task Group were the parties to the Supreme Court Decree, the Delaware river master, and Delaware River Basin Commission staff. The Task Group's work would find its way into the Level B plan and the Good Faith negotiations. After developing a report in March 1979, the Task Group began functioning directly as an advisory committee to the Good Faith negotiations.[15]

The drought criteria developed by the Task Group were contained in a "rule curve" that defined non-drought, drought-warning, and drought emergency conditions as a function of the amount of water stored in the three New York City reservoirs. "Normal" reservoir drawdowns during the summer months were taken into account so that droughts were not declared prematurely. When the storage in the reservoirs dropped into the "drought-warning" portion of the rule curve, the Montague flow requirement was reduced and the city and New Jersey had to reduce their water diversions from the Delaware. If the reservoir levels dropped into the "drought" portion of the rule curve, the flow requirement and diversions were reduced even further. The Task Group did not finalize the exact Montague flow and water diversion reductions, pending the outcome of the Good Faith negotiations.

The Task Group completed its work in 1979, the second wettest year ever in the Delaware River Basin. In January 1979 a three-inch rain caused flooding on the Schuylkill and Lehigh rivers, and it nearly caused flooding on the Delaware. By the end of the year, however, the water picture appeared to be deteriorating. Extra-heavy rain and snow occurred, however, and the Delaware River Basin headed into the summer of 1980 with its major reservoirs nearly full.

The summer of 1980 was noted for its sunny, rainless weekends. On the Upper Delaware Scenic and Recreational River, a record number of visitors was recorded. Meanwhile, the lack of rain was destroying half of Pennsylvania's corn crop, while reservoir levels in the Delaware River Basin and neighboring regions began falling drastically. The first to feel the full force of the drought were the reservoirs serving North Jersey. In September Governor Byrne of New Jersey instituted mandatory water-use restrictions in North Jersey, with stiff fines and jail sentences for violators. In the Delaware River Basin, the first to feel the effects were persons and towns relying on groundwater. Private wells began drying up, and officials who watched over municipal well systems began calling for voluntary bans on nonessential water use.

The Delaware River Basin Commission began taking steps concerning the as yet to be proven drought. In late summer DRBC began ordering releases from Beltzville and Blue Marsh reservoirs in order to augment the freshwater flows reaching the tidal Delaware River. By mid-October the water levels in the New York City reservoirs were down 66 percent and still falling. The criteria developed by the drought Task Group indicated that a drought-warning situation existed, and DRBC declared one on October 17, 1980. Then, using a formula agreed upon by the Good Faith parties, the allowable water diversion of New York City was reduced from 800 mgd to 600 mgd and New Jersey's from 100 mgd to 65 mgd. At the same time, the Montague flow requirement was reduced from 1,750 to 1,655 cfs. With these steps, the Delaware River Basin hoped to save water until precipitation improved reservoir levels.

The hoped-for precipitation did not occur. By early 1981 water levels throughout the Delaware River Basin and northern New Jersey were reaching critical levels. Eighty communities in Pennsylvania were within a month of running out of water, and an estimated four thousand wells serving individual homes had gone dry. Some smaller communities in Pennsylvania were pumping water from abandoned mines for home use. Higher than normal salinity levels were coming up the Delaware Estuary.

On January 15 DRBC met in Trenton to take additional action on the drought. Attending, as commissioners, were the governors of the four Delaware River Basin states and, as advisors, the mayors of New York City and Philadelphia. It was the first DRBC summit meeting since the 1975 Tocks Island vote. This time there was no controversy. A drought emergency was declared, and further reductions were made to the out-of-basin water diversions and the Montague flow requirement. These and other drought management steps taken by DRBC and its states saved more than sixty billion gallons of water.[16]

As the drought deepened, Tocks Island Dam was again being debated. The drought was not the initial reason for the new debate, however. Coming out of the Level B and Good Faith activities was renewed interest in building Hackettstown Dam in New Jersey. New Jersey enjoyed the benefits of the dams located in the Pennsylvania and New York portions of the basin, yet it had no dams of its own. Through the Good Faith talks, Pennsylvania and New York were pressuring New Jersey to build the Hackettstown project as a gesture of New Jersey's good faith.

The Hackettstown project on the Musconetcong River had been kicking around ever since it was first proposed in the Corps' *Delaware River Basin Report*. A 1969 bond issue had authorized $27 million for land acquisition in the planned reservoir area, and by 1979 40 percent of the reservoir lands had been acquired (as Allamuchy State Park). The water-supply phase of the project, however, was still considered to be thirty to forty years down the road. As a result of the pressure from the other states, New Jersey decided to advance the timing of the project. The Hackettstown proposal reopened the Tocks Island Dam controversy because one alternative to building Hacketts-town Dam was to build Tocks.

Among the first to link the construction of Hackettstown Dam to Tocks Island Dam were the *Newark Star Ledger* and Tocks advocates in the New Jersey legislature. The worsening drought brought additional clamoring for an evaluation of New Jersey's Tocks Island Dam decision. The Byrne administration was steadfast, however. At the January 1981 Delaware River Basin Commission summit meeting, Byrne queried his fellow governors and found no support among them for reexamining Tocks Island Dam. Pennsylvania's new governor, Dick Thornburgh, in particular, had moved his state away from the hard pro-Tocks stand of former governor Shapp.

In Pennsylvania the movement to revive Tocks Island Dam was begun by the Philadelphia Chamber of Commerce and other business interests. It went nowhere. On February 15, 1981, however, the *Philadelphia Bulletin* published a persuasive pro-Tocks article entitled "Will Those Who Opposed the Tocks Island Dam (including Gov. Byrne) Please Stand Up!" The author of the article was Frank Dressler, the former executive director of the WRA/DRB and, later, TIRAC. Dressler blamed the demise of Tocks Island Dam on "radical environmentalists," who "did a very skillful political number on Tocks Island." According to Dressler, if the governors changed their minds, Tocks Island Dam was ready to go since the engineering was done and the land already acquired.

The environmentalists were slow to respond to the renewed interest in Tocks Island Dam. In March, however, Barry Allen, the president of the Delaware Valley Conservation Association and vice chairman of the Save the Delaware Coalition, responded to the attempted revival. If Dressler thought that the dam opponents were radical environmentalists, Allen found the "few special interest groups" lobbying for the dam to be an "unholy alliance of utility and union lobbyists."[17] The intensity of the feelings concerning Tocks Island Dam had not abated in the six years since the project's deferral.

During February 1981 seven inches of rain fell on the Delaware River Basin, adding one hundred billion gallons to the basin's reservoirs. The same rains sent water spilling into the North Jersey reservoirs and filled them. The drought emergency was over, and so was the renewed interest in Tocks Island. The drought had been a scare, but the management efforts of DRBC and its members had shown that the basin could cope with a periodic drought of short duration. (This was proven again in the short drought of 1984–85.)

The Hackettstown Dam project, which had initially revived the Tocks Island Dam controversy, had a short life, too. In July 1981 the New Jersey Department of Environmental Protection completed test borings at the dam site and found serious geologic problems. The dam project was then dropped from further consideration.

With the drought over, the Good Faith negotiators wound up their work. Quipped the *New York Times*, "It may have taken the Lord to part the Red Sea, but only four governors and a mayor were required to divide the Delaware River."[18] Following public meetings, the *Interstate Water Management Recommendations of the Parties to the U.S. Supreme Court Decree of 1954 to the Delaware River Basin Commission, Pursuant to Commission Resolution 78-20* began circulating among the four state governors and the mayor of New York City. By early 1983 all had signed the document, and it was formally submitted to the Delaware River Basin Commission for approval.

The Good Faith agreement was as much an attempt to correct the past as it was a plan for the future. In the Good Faith document were recommendations that modernized the provisions of the Supreme Court Decree, the recommendations of the Corps' *Delaware River Basin Report*, and some of the decisions made in the twenty years of the Delaware River Basin Commission. The deferral of Tocks Island Dam had been the major factor initiating the talks, and the environmentalists had wanted alternatives to the dam. The Good Faith recommendations were the response.

The final agreement consisted of fourteen interrelated recommendations:

1. Changes to DRBC's 1967 salinity standards were recommended. In essence the standards were relaxed slightly because the existing and anticipated Delaware River Basin reservoirs did not have sufficient capacity to maintain the old chloride standards during drought. However, a new reference point for the standards was recommended to provide protection to the Camden well fields. In addition to the new chloride standard, a standard for sodium was recommended.

2. A drought equal in magnitude to the 1960s drought was recommended as the underlying assumption for all future water-supply planning in the Delaware River Basin. This recommendation, in essence, declared the policy that the Delaware River Basin would protect itself from such a drought even if the risk of such a drought was low. (The Corps' 1982 salinity report, for example, indicated that the probability that a chloride concentration of 250 mg/l would occur at the mouth of the Schuylkill in any given year was less than 4 percent.)[19]

3. The 1980–81 drought had proven the value of having a formula for reducing reservoir releases and water diversions when drought or drought-like conditions appeared in the basin. The criteria developed by the drought Task Group were recommended for adoption. Based on the criteria, out-of-basin water diversions could be reduced as much as 32 percent. The amount of water flowing down the the Delaware River would also be decreased by reducing the Montague flow requirement and the Trenton flow objective. Significantly, the flow objectives were reduced percentage-wise to a lesser extent than the out-of-basin diversions. In other words, the drought management criteria favored in-basin water needs. This provision represented a major concession of New York City to the Good Faith agreement.

4. The development of a plan for coordinating the operation of the New York City reservoirs with the other reservoirs in the Delaware River Basin during drought periods was recommended. This recommendation assumed that DRBC would take jurisdiction over the operations of federal, state, and power company reservoirs during a drought. It had done so twice in the past.

5. Dam projects were recommended. By 1990 the F. E. Walter flood-control reservoir was to be enlarged for water-supply storage, as was Prompton Reservoir by 1995. The Merrill Creek high-flow skimming project was recommended for construction as early as 1986. These three projects added 620 cfs of new flow-augmentation ability to the Delaware River Basin.

The yield of the three recommended projects was important. The Level B study had determined that the existing reservoirs in the basin could maintain a minimum flow at Trenton of 2,500 cfs. By the year 2000, however, increased depletive use was projected to reduce this flow by 200 cfs. DRBC's salinity model, however, had predicted a need for a Trenton flow of 2,900 cfs to meet the Good Faith salinity objectives. Thus, a shortfall of 600 cfs would occur, even if water conservation measures were taken.

6. The enlargement of New York City's Cannonsville Reservoir was also recommended. The enlargement was to be completed by the State of New York by 1990. The extra water was intended to provide drought protection for the cold-water fishery in the upper basin.

7. By 1985 the State of New Jersey was to develop a plan for solving the groundwater problems of the Camden metropolitan area. The plan was to be implemented by 1990. The purpose of this recommendation was to reduce South Jersey's reliance on the vulnerable P-R-M Aquifer.

8. Groundwater located in alluvial deposits in the upper basin was to be examined as a potential emergency source of water for use after the year 2000. This recommendation, highly controversial in the upper basin, was based on the idea that groundwater could be used to augment river flows during drought. By pumping the groundwater into the river, it was made available for downstream areas, which presumably needed it most.

9. Tocks Island Dam was not to be deauthorized but held in reserve for development after the year 2000—if needed for water supply. The project's description in the DRBC Comprehensive Plan was to be modified to reflect the changed status of the project.

10. The amount of water in New York City's Delaware River Basin reservoirs was to be the principal consideration for declaring and suspending drought emergencies. The recommendation established the criteria of recommendation 3 as policy and outlined the method in which it would be used.

11. A goal of reducing depletive water use by 15 percent was recommended for adoption as a DRBC policy. The 15 percent figure had been derived from the Level B study, and reductions in water use during the 1980–81 drought had verified that a 15 percent reduction was attainable.

12. Each state was to develop a drought contingency plan describing how it was going to obtain the desired reductions.

13. DRBC was to develop a program to regulate new depletive water uses. Increased depletive water use could offset any gains in water storage brought about by the agreement. This was one of the most important recommendations since it recognized that the amount of reservoir storage

that can be built is limited. (This program will break new ground in the water field.)

14. The last recommendation made the experimental conservation releases program for the New York City reservoirs permanent. Studies during the experimental period had shown that such a program had a beneficial impact on cold-water fish. Unlike the other Good Faith recommendations, this one is of minor importance to the Delaware River Basin as a whole.

The first Good Faith recommendations to be adopted by DRBC were those dealing with the salinity standards and drought operations. Public hearings were held, and in June 1983 the Delaware River Basin Commission adopted the new salinity standards, the drought management criteria, and the water conservation goals. There was little opposition to these. The next round of Good Faith recommendations dealt with the recommended reservoir projects. In August 1983 DRBC held hearings in the affected areas. There was local opposition to each project. Significantly, many of the environmentalists who had opposed Tocks Island Dam supported the Good Faith dam projects or at least remained neutral.

Tocks Island Dam, the object of intense controversy over the years, was the subject of relatively subdued testimony. Opponents of the dam were not pleased to find the project kept alive until the year 2000, and many of the environmental arguments were raised again. On the other side, Tocks Island supporters wanted the dam delayed only until 1990. They predicted dire consequences if Tocks was delayed too long. The position of the New Jersey Section of the American Society of Civil Engineers, for example, was printed verbatim in the July 7, 1983, issue of *Engineering News-Record* under the title "Water Crisis Coming in the East." It talked about "the peril of water shortages," "catastrophe," and "irreversible damage" that might occur if Tocks Island Dam was not built. The article accused the Good Faith participants of being "very casual about saline contamination."

DRBC action on the three proposed reservoir projects and the deferral of Tocks Island Dam came on November 30, 1983. Resolution 83-27 modified the description of the Tocks Island Dam Project that had appeared in the DRBC Comprehensive Plan since 1962. Added also was a clause that called for a complete review of the environmental impacts of Tocks Island Dam before the project was ever reactivated. The additions had been suggested during the hearing process by the anti-Tockers.

Following the adoption of the Good Faith recommendations, DRBC's staff and the staff of the four basin states began working on their assigned tasks. State drought contingency plans were developed on schedule, although ques-

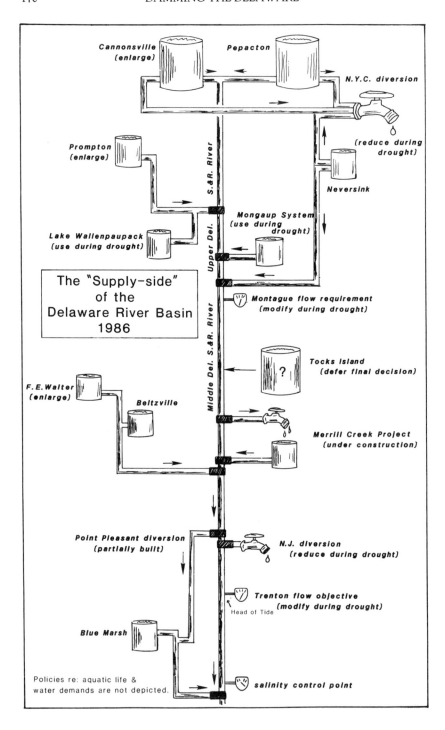

tions remain about their completeness. New Jersey initiated its required study of the South Jersey aquifers. In early 1986 the state announced that the southern half of the Potomac-Raritan-Magothy Aquifer would be declared a critical area and managed accordingly. A mandatory cut in water use of 35 percent may be instituted. Meanwhile, the state is evaluating alternative sources of water. Water piped across the Delaware from Philadelphia appears to be the cheapest alternative, but there is reluctance to go this route.

In Philadelphia the Corps of Engineers is once again working on dam projects in the Delaware River Basin—the expansions to Walter and Prompton dams. The design work and environmental impact assessment process are under way. In Trenton the DRBC is attempting to institute a water charge system to pay for the reservoir expansions. It has asked Congress to modify Section 15.1(b) of the compact, which exempts water use that existed prior to the adoption of the compact from the commission's water charges. The proposal to change the federal compact language recognizes that drought protection serves all users, not just new ones. Meanwhile, DRBC staff are attempting to develop depletive water use budgets and other materials related to the Good Faith agreement.

In 1985 the Good Faith drought criteria were used once again. A water shortage occurred in August 1984, and by January 1985 a drought warning was declared. The situation continued to deteriorate, however, and a drought emergency was declared in May. By this time groundwater levels had fallen drastically throughout the Delaware River Basin, and water managers were extremely concerned about the water levels in their reservoirs. Owing to the lack of rain and the cutbacks in reservoir releases, flows in the Delaware reached record lows for the month of April. Gradually, however, the situation improved, culminating with the rains of Tropical Storm Gloria in late September. By the end of 1985, the drought restrictions had been lifted. It was a close call, nevertheless.

The cry to build Tocks Island Dam was rarely heard during the severe but short 1984–85 drought. About the only official to publicly mention the dam project was the commander of the Corps' North Atlantic Division, Brig. Gen. Paul Kavanaugh. Kavanaugh stated that the dam's construction would have averted the drought restrictions placed on North Jersey. Privately, many pro-Tockers were thinking the same thing.

Congress may have placed the last nail in the Tocks Island Dam coffin. On October 17, 1986, it enacted the Water Resource Development Act of 1986 (P.L. 99-662), which reforms and modernizes the Corps of Engineers' public works program. Under Section 203 of the act, Tocks Island Dam (and other inactive projects) will automatically be deauthorized on October 17, 1987. If Tocks survives this deauthorization or is reauthorized at a future date, it will

Epilogue

In October 1983 I drove to Warren County to see the dam. It is located four miles from the main highway on a narrow road that parallels the river. After a sufficient number of bumps and jolts on this road, I could see the dam stretching 1,900 feet across the valley. I drove past the visitors center to the dam overlook, with its American and Corps of Engineers flags. There one could see the crest of the dam. It rises 179 feet above the riverbed and dominates the scenery of the whole valley. In the center of the dam, four large gates control the discharge from the reservoir.

At the base of the dam are a national fish hatchery and a power station. Turbulence behind the dam indicated that the dam's pumped-storage project, owned by two large power companies, was generating electricity on the day of my visit. Above the dam stretched the blue-gray expanse of the reservoir. Its ninety miles of shoreline contain sixteen camping areas, four swimming beaches, eight picnic areas, eight boat launches, and two marinas. A variety of fish await the fisherman. In the recreation area surrounding the reservoir are 168 miles of hiking trails, off-road bike trails, backpacking and hunting areas, and a variety of other outdoor opportunities.

My visit was not imaginary. I was in Warren County, Pennsylvania, and not Warren County, New Jersey, the home of Tocks Island Dam. The dam I visited is Kinzua Dam, the pride of the Pittsburgh District, U.S. Army Corps of Engineers. Surrounding the reservoir is the Allegheny National Forest, consisting of a half-million acres of land. If a person wants to get a feel for what Tocks Island Reservoir and the Delaware Water Gap National Recreation Area might have looked like, this is the place to go.

I had traveled to see "Kinzua Country" because of its similarity to the Tocks Island Dam Project. I am quite familiar with the undammed Delaware River that flows past Tocks Island and its infamous dam site. I was curious to see Kinzua Dam and Allegheny Reservoir. Was Kinzua some kind of environmental disaster area or a nice place to visit? I liked it, but others might not. Already some people find the undammed Delaware River too crowded for their personal enjoyment. I suspect these people don't go to Kinzua either.

Kinzua Dam is primarily a flood-control dam, and its credentials are impressive. During the June 1972 "Agnes" flood, for example, the dam prevented flood damages that were equivalent to or greater than its original cost. It is difficult to compare the need for Kinzua Dam to the need for Tocks Island Dam, however. Tocks was made economically feasible by allowing recreation benefits to be used in its benefit-cost ratio. In the opinion of many, recreation is not comparable to flood control and water supply as a water-resource purpose. If water-supply and flood-control benefits had been large enough to justify the construction of Tocks in themselves, the dam probably should have been built. Still, one large flood or one severe drought can go a long way toward paying for an expensive dam project. Kinzua demonstrates this.

A forceful argument can be made that the Delaware River should never be dammed. A free-flowing, natural river (albeit controlled) is a beautiful thing that should be preserved for future generations. The designation of two reaches of the Delaware as national scenic rivers recognizes this. These designations will be important impediments for future dam builders, and rightfully so. The value of a preserved scenic river will increase over the years. Already, interest in the Delaware River as a regional resource is rising, and it is quite exciting. The new riverfront parks, the many shad festivals, the summer tubing and canoeing, the great fishing, and the other river attractions are building a constituency for a free-flowing river. It is difficult to imagine similar affection for a reservoir.

To dam or not to dam a river is really an emotional issue that does not lend itself to rational discourse. As a result, dam fights tend to polarize the positions on both sides of the controversy. A free-flowing river becomes a sacred thing, and the dam is seen as absolutely necessary. Unfortunately, no middle ground exists between damming a river and leaving it alone. A dam, once built, is irrevocable. Technical arguments will rarely resolve dam disputes because both sides are often right. It depends on one's philosophy and view of the future.

It would be dangerous to predict whether Tocks Island or some other dam will ever be built on the Delaware River, but it is very doubtful. Time has caught up with the idea of a dam across the Delaware. The idea was born in an era when an undammed river was considered a wasted resource and river taming was considered good water conservation. The momentum of this philosophy carried Tocks Island Dam to the start of construction, but it was not enough. Today the era of big dam building is over. The Water Resource Development Act of 1986 demonstrates this.

Many persons and organizations influenced the decision that stopped Tocks Island Dam. It was the Vietnam War, however, that kept the project from being built. In many respects, Tocks Island was the Vietnam War of the Delaware Valley. Tocks had its U.S. Army, spiraling costs, dogmatic supporters, opposition protesters, destroyed villages, and even its own hippies. Its life span paralleled the war, and, like the war, its underlying assumptions had evolved out of decades of controversy. The mid-1950s saw a massive escalation of the federal role in the affairs of both the Delaware and Vietnam. However, as the cost of achieving the underlying goals escalated, they became confused and were questioned. In 1975 both the dam and the war collapsed from weakened political support and rising public pressure.

The most lasting legacy of the Tocks Island Dam Project will be the Delaware Water Gap National Recreation Area. As a national recreation area based on a scenic, free-flowing river, it is unique. Thousands use it for recreation. I have stood on the banks of the Delaware River and watched whole busloads of people enter the river to float downstream on inner tubes, rafts, and canoes. Along its banks, too, hundreds pursue other activities, or sit and watch the river go by. This is my vision of the Delaware River as it should be: a modern-day version of the old-fashioned swimming hole, used by many people for innocent pleasure.

The most recent legacy of Tocks Island Dam is the Good Faith agreement. The agreement marks a change in water-resource philosophy in the Delaware River Basin, one that moves away from an all or nothing belief in a mainstem dam. It is to be hoped that the Good Faith agreement will prove that the Delaware River Basin states can work together to find incremental solutions to common problems. If not, the year 2000 may find Tocks Island Dam again promoted as the only viable solution for solving water problems in the Delaware River Basin states. It would be a shame if that happened.

Notes

CHAPTER 1

1. *World Almanac and Book of Facts, 1981,* pp. 444–45.
2. *Congressional Record—House* (26 July 1976), p. 23900.

CHAPTER 2

1. Alfred Chandler, cited in *Proceedings of the Canal History and Technology Symposium,* vol. 2, p. 113.
2. Carter Goodrich, *Government Promotion of American Canals and Railroads,* p. 123.
3. N.Y. State Water Supply Commission, *Third Annual Report* (1908), p. 299.
4. J. C. Bell, "Official Opinion of the (Pa.) Attorney General," p. 3.

CHAPTER 3

1. George Profus, "The Old Croton Aqueduct," in *The Conservationist* (November 1981), p. 39.
2. Charles H. Weidner, *Water for a City,* p. 58.
3. Incodel, "Chronology of the Water Supply Problems of New York City, Philadelphia, and Northeastern New Jersey," p. 2.
4. N.Y. State Water Supply Commission, *Fourth Annual Report* (1909), p. 24.
5. *New York Times* (13 February 1927), sec. 2, p. 1.
6. Maynard M. Hufschmidt, *The Supreme Court and Interstate Water Problems—The Delaware Basin Example,* p. 28.

7. Robert E. Horton, *A Program for Regulation and Conservation of the Delaware River*.

CHAPTER 4

1. *Rivers and Harbors Act of March 3, 1925*.
2. Barbara T. Andrews and Marie Sansome, *Who Runs the Rivers?* p. 40.
3. U.S. Congress, House Document 522 (*Delaware River 308 Report*), p. 88.
4. Ibid., p. 102.
5. Ibid., Letter of Transmittal of the Division Engineer.
6. U.S. Congress, House Resolution of the Committee on Rivers and Harbors (24 January 1939).
7. U.S. Army, *Preliminary Examination Report, Review of 308 Report on the Delaware River*, p. 51.
8. A. Harry Moore, *Special Message to the 162d Legislature of New Jersey*, p. 11.
9. Incodel, *Annual Report* (1943), p. 51.
10. For example: James H. Allen article in the *Journal American Water Works Association* (January 1944).
11. Board of Consulting Engineers, *Preliminary Report to the Philadelphia Water Commission*.
12. Ibid., p. 16.
13. Board of Consulting Engineers, *Report to the Philadelphia Water Commission*.
14. Harris-Dechant Associates, *Report—Diversion of the Waters of the Delaware River by New York City and Northern New Jersey*, transmittal letter from M. F. Draemel to Governor Duff.

CHAPTER 5

1. *Engineering News-Record* (17 June 1954), p. 152.
2. Samuel Baxter, *Engineering News-Record* (25 June 1954), p. 12.
3. *Engineering News-Record* (29 July 1954), p. 116.
4. Interview with W. Brinton Whitall.
5. Samuel Baxter, letter to Joseph S. Clark (9 August 1954).
6. U.S. Geological Survey, *Floods of August 1955 in Northeastern States*, p. 8.
7. Michael Reich, in *Boundaries of Analysis*, p. 48.

CHAPTER 6

1. U.S. Congress, Senate Resolution (13 April 1950).
2. Vernon D. Northrop, letter to Maynard M. Hufschmidt (28 October 1955).
3. U.S. Congress, Senate Resolution (14 September 1955).
4. Francis A. Pitkin, statement at congressional hearing at Mt. Pocono, Pa. (21 September 1955).
5. U.S. Congress, Senate Resolution (20 February 1956).
6. "Accelerated Study, Tocks Island Site." Letter report from Colonel Allen F. Clark, Jr., to Governor Leader of Pennsylvania, 6 February 1957. In *Delaware River Basin Report*, vol. 2, appendix A. U.S. Army, 1960.

7. Francis A. Pitkin, "Tocks Island in 1958," *Annual Report of Chairman 1957*, p. 7.

8. Charles R. Bensinger, Statement at Senate hearing (30 October 1959).

9. Delaware River Basin Advisory Committee, "Second Directive" (October 1959).

10. *Philadelphia Bulletin* (14 April 1961).

11. William Voight, Jr., *The Susquehanna Compact*, p. 67.

12. Public hearing at Wilmington, Del. (20 October 1958), *Delaware River Basin Report*, vol. 2, appendix A, exhibit B, pp. 28–41.

13. Michael Reich, in *Boundaries of Analysis*, p. 50.

CHAPTER 7

1. U.S. Congress, *Rivers and Harbors Act of 1962*, title 2, sec. 203.

2. Tocks Island Regional Advisory Council, *Memo from TIRAC No. 1* (22 April 1966) and E. P. Yates statement (16 April 1964).

3. U.S. Army, *Design Memorandum No. 1, Site Selection*, chap. 6, pp. 2–3.

4. U.S. Army, *Design Memorandum No. 1, Site Selection, Supplement No. 1*, chap. 1, p. 2.

5. Ibid., chap. 1, p. 4.

6. Delaware River Basin Commission, "Resolution No. 65-24."

7. Clarke and Repuano, *Tocks Island Dam, A Plan for Its Architecture*, p. 6.

8. *Pocono Record* (20 June 1969).

9. "Time Essay: How to Cut the U.S. Budget." *Time* (8 December 1967), p. 38.

10. *Pocono Record* (21 March 1968).

11. URS/Madigan-Praeger, *A Comprehensive Study of the Tocks Island Lake Project and Alternatives*, vol. B, chap. 7, p. 16.

12. Walter J. Sheldon, *Tigers in the Rice*, p. 95.

13. Ibid., p. 337.

14. Ibid., p. 338.

15. Lyndon B. Johnson, "The Budget Message," *Weekly Compilation of Presidential Documents* (31 January 1966), p. 82.

16. Lyndon B. Johnson, "The President's News Conference of January 17, 1967," *Weekly Compilation of Presidential Documents* (23 January 1967), p. 54.

17. Lyndon B. Johnson, "The Budget Message," *Weekly Compilation of Presidential Documents* (5 February 1968), p. 164.

18. Ibid., p. 152.

19. Ibid., p. 158.

20. James Lawrence, *Trenton Evening Times* (27 January 1970).

21. Council on Environmental Quality, *The First Annual Report*, pp. 292–93.

CHAPTER 8

1. Anthony J. Francoline, "The Tocks Island National Recreation Area," p. 2.

2. U.S. Army, *Delaware River Basin Report*, vol. 4, appendix 1, p. 122.

3. Ibid., p. 132.

4. Ibid., p. 127.

5. Charles Bensinger, Statement in Senate hearings (30 October 1959).

6. Water Resources Foundation for the Delaware River Basin, *Water for Recreation, Today and Tomorrow*, p. 21.

7. Ibid.

8. James Kerney, Jr., Statement in U.S. Congress, *Hearings on H.R. 89*, p. 70.

9. *Times Herald Record* (25 October 1963).

10. Francoline, "Tocks Island," p. 16.

11. National Park Service, *Tocks Island National Recreation Area—A Proposal*, pp. 4–5.

12. *New Jersey Herald* (3 April 1966).

13. URS/Madigan-Praeger, *A Comprehensive Study of the Tocks Island Lake Project and Alternatives*, vol. B, chap. 7, p. 16.

14. Robert R. Nathan and Associates, *A Central Park in Megalopolis*, p. 81.

15. Raymond and May, *Preface to Planning*, p. 11.

16. Nathan, "Central Park," p. 104.

17. "Pennsylvania to Improve Its Roads; Jersey Has No Plans," *Trenton Evening Times* (12 October 1967).

18. Edwards and Kelcey, Inc., *Approach Roads Study, Tocks Island Region, Part II*, p. 90.

19. Roy F. Weston, *Tocks Island Region Environmental Study*, vol. 1, p. 2.

20. Ibid.

21. Ibid., p. 118.

22. Ibid., p. 2.

23. Ibid., p. 118.

CHAPTER 9

1. Lenni Lenape League, letter (9 August 1967).

2. Douglas W. Smith, *New Jersey Outdoors* (January 1985).

3. Robert Meyner, statement at DRBC hearing (17 August 1967), p. 85.

4. Richard Gross, conversation with author.

5. Lenni Lenape League, "The Case for Sunfish Pond," p. 2.

6. Jersey Central Power and Light, *Yards Creek Pumped Storage Generating Station*, p. 2.

7. *Pocono Record* (21 April 1956), p. 1.

8. U.S. Army, *Delaware River Basin Report*, vol. 10, appendix T, pp. 41, 45.

9. Public hearing at Phillipsburg, N.J. (13 April 1960), *Delaware River Basin Report*, vol. 2, appendix A, exhibit B, pp. 44–45.

10. Fredrick W. Heilich III, *The History of the Blairstown Railway*, p. 45.

11. DRBC, "Docket No. D-62-2," in Minutes of 29 August 1962 meeting.

12. National Park Service, *The Master Plan, DWGNRA*, p. 4.

13. Lenni Lenape League, letter (9 August 1967) and Frank Thompson, Jr., press release (1 June 1967).

14. *Trenton Sunday Advertiser* (18 June 1967).

15. *Netcong News Leader* (8 August 1968).

16. DRBC, "Resolution No. 68-12."

17. Glenn Fisher, letter in *Easton Express* (20 August 1968).

18. *New Hope News* (14 May 1970).

CHAPTER 10

1. Russell E. Train, letter to Robert E. Jordan (7 April 1971), *Congressional Record—House* (28 July 1971), p. 27792.

2. Ibid.

3. U.S. Army, *Environmental Impact Statement—Tocks Island Lake,* chap. 1, p. 4.

4. *Congressional Record—House* (29 July 1971), p. 28096.

5. *Congressional Record—House* (27 July 1971), p. 27791.

6. *Congressional Record—House* (29 July 1971), p. 28096.

7. *Phillipsburg Free Press* (18 June 1972).

8. *Congressional Record—House* (29 July 1971), p. 28097.

9. *Easton Express* (12 December 1966).

10. *Pocono Record* (12 December 1966).

11. Thomas H. Cahill, "Cultural Eutrophication in Tocks Island Reservoir," p. 66.

12. Ibid., p. 67.

13. Thomas H. Cahill, telephone interview with author.

14. Ralph Manning, *The Poultry Industry in the Upper Delaware River Basin.*

15. Jack McCormick and Associates, *An Appraisal of the Potential for Cultural Eutrophication of Tocks Island Lake,* p. 107.

16. Russell E. Train, letter to Robert F. Froehlke (21 October 1971), in *Congressional Record—Senate* (2 February 1972), pp. 2338–39.

17. *Newark Evening News* (28 August 1972).

18. DRBC, "Resolution No. 70-4."

19. *Newark Evening News* (28 August 1972).

20. William T. Cahill, letter in *Trenton Evening Times* (28 September 1972).

21. *New York Times* (15 September 1972), p. 77.

CHAPTER 11

1. Nancy Skukaitis, interview with author.

2. Ibid.

3. *New Jersey Herald* (24 March 1971).

4. Ralph Abele, statement in *Tocks Island Deauthorization Hearings,* p. 304.

5. For example: Gretchen Leahy, statement at DRBC Tocks Island Public Information Meeting (22 May 1975).

6. U.S. Army, *Environmental Impact Statement—Tocks Island Lake,* p. c-1.

7. Harold A. Feiveson, in *Boundaries of Analysis,* p. 85.

8. Ibid., p. 91.

9. George E. Schindler and Frank W. Sinden, eds., *Papers in Support of a Free-Flowing River,* p. xii.

10. Feiveson, p. 91.

11. DRBC, *Annual Report 1970,* p. 19.

12. DRBC, *Water Demands in the Delaware River Basin as Related to the Tocks Island Reservoir Project,* p. 33.

13. Allan S. Krass, in *Boundaries of Analysis,* p. 222.

14. Environmental Defense Fund, *Flood Control and the Delaware River,* 1973, pp. 65–70.

15. U.S. Congress, *A Legislative History of the Water Pollution Control Amendments of 1972,* p. 241.

16. DRBC, "Resolution No. 72-11."
17. DRBC, "Resolution No. 73-5."
18. DRBC, "Resolution No. 73-6."
19. *New York Times* (1 June 1973), p. 2.
20. WRA/DRB, *Memo from WRA/DRB* (1 September 1973).
21. *Congressional Record—House* (28 June 1973), p. 22112.
22. Malcolm Wilson, letter to Joe Evins (2 May 1974).
23. Frank Thompson, Jr., letter to Joe Evins (1 March 1974).
24. *Easton Express* (11 October 1974).

CHAPTER 12

1. Phillip Morrissey, statement at House hearings (5 June 1969); *Minisink Bull* (31 July 1968), p. 7; Ruth Jones, letter to author; *Newark Evening News* (19 May 1969); and *Minisink Bull* (8 September 1967), p. 7.
2. General Accounting Office, *Review of the Tocks Island Reservoir Project*, p. 27.
3. Rod Nordland, *Philadelphia Inquirer* (2 June 1974), pp. 2–4.
4. *Pocono Record* (17 November 1972).
5. *Pocono Record* (13 November 1972).
6. William Read, interview with author.
7. Dorothy Belmont, interview with author.
8. Ibid.
9. *Pocono Record* (21 November 1972).
10. James M. Markham, *New York Times* (6 September 1971), p. 22.
11. James M. Markham, *New York Times* (7 September 1971), p. 31.
12. *Pocono Record* (12 June 1973).
13. Read and Belmont, interviews.
14. *Newark Star-Ledger* (20 February 1974).
15. *Pocono Record* (6 March 1974).

CHAPTER 13

1. Mina Haefele, statement at DRBC Tocks Island Public Information Meeting (22 May 1975).
2. DRBC, transcript of press conference (31 July 1975).
3. URS/Madigan-Praeger, *A Comprehensive Study of the Tocks Island Lake Project and Alternatives*, vol. D, chap. 14.
4. *New York Times* (5 August 1975), p. 65.
5. U.S. Congress, *Public Law 90-545*, sec. 1 (b).
6. *Newark Sunday News* (3 March 1968), p. 35.
7. *Congressional Record—House* (10 July 1978), p. 19965.
8. Ibid., p. 19966.
9. Ibid., p. 19967.

CHAPTER 14

1. *New York Times* (12 October 1974), p. 67.
2. Letter from Department of the Interior (Herbst) to Representative Udall refer-

enced in Milton J. Shapp letters (22 June 1978) to Governors Byrne (N.J.) and Carey (N.Y.).

3. Shapp, letters to Byrne and Carey (22 June 1978).

4. DRBC, Minutes of 25 October 1978 meeting, p. 2.

5. DRBC, "Resolution No. 78-20."

6. DRBC, "Resolution No. 71-3."

7. Incodel, A Report on Its Activities and Accomplishments, 1943, p. 19.

8. Ibid., p. 32.

9. U.S. Army, Information Bulletin, Delaware River Basin Study, pp. 14, 30.

10. U.S. Army, Delaware Estuary Salinity Intrusion Study, p. 18.

11. James F. Wright, letter to General Kelley, in U.S. Army, Delaware Estuary Salinity Intrusion Study, appendix 4, p. 8.

12. Resolution of House Committee on Public Works and Transportation, in U.S. Army, Delaware Estuary Salinity Intrusion Study.

13. Delaware River Basin Commission, Delaware River Basin Comprehensive Study—Draft, 1979.

14. Delaware River Basin Commission, Docket No. D-77-20.

15. Delaware River Basin Commission, Task Group Report, 1979.

16. Delaware River Basin Commission, Annual Report, 1982, p. 6.

17. Delaware Valley Conservation Association, press release, 18 March 1981.

18. New York Times (27 February 1982).

19. U.S. Army, Delaware Estuary Salinity Study, p. 18.

Sources

The following sources were augmented by use of the news-clipping files of the Delaware River Basin Commission (Public Information Office) and by extensive research in the *New York Times*, the *Pocono Record* (and its predecessor, the *Stroudsburg Record*), and the *Minisink Bull.*

Able, Ralph. Testimony in *Tocks Island Deauthorization Hearings Before the Subcommittee on Water Resources, July 23 and 26.* 95th Cong., 2d sess., 1978.
"Accelerated Study, Tocks Island Site." Letter report from Colonel Allen F. Clark, Jr., to Governor Leader of Pennsylvania, 6 February 1957. In *Delaware River Basin Report.* Vol. 2, Appendix A. U.S. Army, 1960.
Albright and Friel, Inc. *Report to the Commonwealth of Pennsylvania Department of Forests and Waters on Wallpack Bend Dam and Reservoir on the Delaware River at Bushkill, Pennsylvania.* Philadelphia, 1955.
Allen, James H. "The Delaware River Basin: A Home Rule Program for the Development of Its Resources." *Journal American Water Works Association* (January 1944). Reprinted by Incodel.
Andrews, Barbara T., and Marie Sansone. *Who Runs the Rivers? Dams and Decisions in the New West.* Stanford, Calif.: Stanford Environmental Law Society, 1983.
Barksdale, Henry C., David W. Greenman, Solomon M. Lang, George S. Hilton, and Donald E. Outlaw. *Ground Water Resources in the Tri-State Region Adjacent to the Lower Delaware River.* Prepared by the U.S. Geological Survey. Published as Special Report 13 by the New Jersey Dept. of Conservation and Economic Development, Division of Water Policy and Supply. 1958.
Baxter, Samuel. Letter to Joseph S. Clark, Mayor of Philadelphia, 9 August 1954.
Bell, J. C. "Official Opinion of the Attorney General." Harrisburg, Pa., 1911.
Belmont, Dorothy. Interview with author. East Stroudsburg, Pa., 1985.

Bensinger, Charles R. Statement before the Senate Select Committee on Natural Resources, 30 October 1959.

Board of Consulting Engineers. *Preliminary Report to the Philadelphia Water Commission: A New Water Supply from Upland Sources.* Philadelphia, 1945.

———. *Report to the Philadelphia Water Commission: Development of an Upland Source of Water Supply and Suitability of Existing Sources of Supply with Augmented Facilities.* Philadelphia, 1946.

———. *Report to the Philadelphia Water Commission on Review of Lehigh Coal and Navigation Company Plan for a Water Supply from the Upper Lehigh for the City of Philadelphia.* Philadelphia, 1946.

Cahill, Thomas H. "Cultural Eutrophication in Tocks Island Reservoir." Master's thesis, Villanova University, 1968.

———. Telephone interview with author, 1983.

Candeub, Fleissig and Associates. *A Concept Plan for the Delaware River.* Prepared for the Save the Delaware Coalition. 1974.

Chandler, Alfred. "Anthracite Coal and the Beginnings of the Industrial Revolution." *Harvard Business History Review* 46, no. 22 (1972). Cited in Sayenga, Donald. "The Untryed Business: An Appreciation of White and Hazard." In *Proceedings of the Canal History and Technology Symposium,* vol. 2, 113. Easton, Pa.: The Center for Canal History and Technology, 1983.

Clarke and Repuano, Inc. *Tocks Island Dam: A Plan for Its Architecture and Development.* Prepared for the U.S. Army Corps of Engineers. New York, ca. 1974.

Compact as to the Waters of the Delaware River. Prepared by New Jersey, New York, and Pennsylvania. 1925.

Compact as to the Waters of the Delaware River. Prepared by New Jersey, New York, and Pennsylvania. 1927.

Congressional Record. Senate. Letter from Russell E. Train to Robert F. Froehlke. 2 February 1971: 2338–39.

———. Remarks of Mr. DuPont offering an amendment to the Public Works Appropriation Bill and 7 April 1971 letter from Russell E. Train to Robert E. Jordan. 27 July 1971: 27791–93.

———. House. Debate Between Mr. DuPont and Mr. Thompson. 29 July 1971: 28096–97.

———. Senate. Remarks of Senator Case. 22 September 1971: 14723–24.

———. House. Statement of Mr. Rhodes concerning the impact of P.L. 92-500 on Tocks Island. 28 June 1972: 22112.

———. House. Letter from Major Gen. R. H. Groves to Frank Thompson, Jr., of 12 June 1972. 26 June 1972: 22431–32.

———. House. Debates on the designation of the Middle Delaware Scenic and Recreation River and the amendment of Mr. Thompson. 10 July 1978: 19959–76.

Council on Environmental Quality. "Statement on Proposed Federal Actions Affecting the Environment, Interim Guidelines." 30 April 1970. In *The First Annual Report of the Council on Environmental Quality.* Washington, D.C.: GPO, 1970.

Delaware River Basin Advisory Committee. "Minutes of 42 Meetings Held between 17 February 1956 and 23 May 1962."

———. "Second Directive to the Delaware River Basin Advisory Committee."

Signed by Governors Boggs (Del.), Rockefeller (N.Y.), Meyner (N.J.), and Lawrence (Pa.), and Mayors Dilworth (Philadelphia) and Wagner (New York City). October 1959.

Delaware River Basin Commission. *Annual Report.* 1963 through 1985.

———. "Background Paper: Financing Basin Water Projects and Potential Modification of Section 15.1 (b) of the Delaware River Basin Compact." 1985.

———. "Background Report Concerning the Interstate Water Management Recommendations." 1982.

———. *Delaware River Basin Comprehensive Study-Draft.* 1979.

———. *The Delaware River Basin: The Final Report and Environmental Impact Statement of the Level B Study.* 1981.

———. *First Phase Comprehensive Plan.* Philadelphia, 1962.

———. Minutes of Meetings.

———. "Resolution No. 65-24: A Resolution Requesting Water Supply Storage at the Tocks Island Project and to Provide Reasonable Assurance to the U.S. Army as to Payments Therefor." 13 September 1965.

———. "Resolution No. 68-9: A Resolution with Respect to Water Supply Storage at the Tocks Island Project Amending No. 65-24." 25 September 1968.

———. "Resolution 68-12: Amending the Comprehensive Plan with Respect to Generation of Hydroelectric Power at Tocks Island Reservoir." 22 October 1968.

———. "Resolution 70-4: In Support of Construction of the Tocks Island Dam and Reservoir." 14 April 1970.

———. "Resolution 71-1: A Resolution Relating to the Selection of Sites for Future Major Electric Generating Stations." 7 April 1971.

———. "Resolution 72-11: A Resolution Relevant to Water Quality Protection in Tocks Island Reservoir." 8 November 1972.

———. "Resolution 72-2: A Resolution to Amend the Comprehensive Plan in Relation to the Protection of Water Quality in the Tocks Island Area." 26 January 1972.

———. "Resolution 73-5: A Resolution to Amend the Comprehensive Plan in Relation to the Protection of Water Quality in the Tocks Island Area." 31 May 1973.

———. "Resolution 73-6: A Resolution Amending the Comprehensive Plan with Regard to Recreation Developments at the Proposed Tocks Island Reservoir and the DWGNRA." 31 May 1973.

———. "Resolution 74-12: Requesting U.S. Congress for Land Acquisition Funds and a Final Determination on Tocks Island and the DWGNRA." 31 July 1974.

———. "Resolution 78-20: Inviting the Parties to the 1954 U.S. Supreme Court Decree to Enter into Good Faith Discussions for Management of the Waters of the Delaware Basin." 13 December 1978.

———. "Resolution 83-27: To Amend the Comprehensive Plan to Revise and Update the Description of the Tocks Island Project." 30 November 1983.

———. *Task Group Report: DRBC Docket No. D-77-20 Appraisal of Upper Basin Reservoir Systems, Drought Emergency Criteria and Conservation Measures.* 1979.

———. Transcript of Proceedings in the Matter of Upper Delaware Estuary Salinity Seminar. Held in Newark, N.J., 17 July 1975.

————. Transcript of DRBC Press Conference. 31 July 1975.

————. *Water Demands in the Delaware River Basin as Related to the Tocks Island Reservoir Project.* 1971.

Delaware River Basin Compact. 1961.

Delaware Valley Conservation Association. "Complaint: DVCA versus Resor, Udall and Cassiday." Civil Action File 9675, U.S. District Court for the Middle District of Pennsylvania. 2 November 1966.

————. Press release from Barry Allen. 18 March 1981.

Delaware Valley Council. *Questions and Answers about the Tocks Island Dam and Reservoir Project.* Philadelphia, 1972.

————. *In-Basin Availability of Up to 100 Million Gallons a Day of Delaware River Water Pending New Jersey Performance of "Specified Condition" for Its Trans-Basin Diversion (Research Memorandum).* 1978.

Detweiler, Thomas. Interview with author. Stroudsburg, Pa., 1981.

Edwards and Kelcey, Inc. *Approach Roads Study, Tocks Island Region, Part I: The Roadway Master Plan.* Prepared for the New Jersey Dept. of Transportation. Newark, 1969.

————. *Approach Roads Study, Part II: Roadway Feasibility Studies.* 1971.

Engineering News-Record. Editorial (17 June 1954): 152.

Engineering News-Record. Editorial (29 July 1954): 116.

Environmental Defense Fund. *Evaluation of the Proposed Tocks Island Reservoir Project.* East Setauket, N.Y., 1972.

————. *Flood Control and the Delaware River.* 1973.

————. *Questions Surrounding Designation of the Middle Delaware as a Wild and Scenic River.* Washington, D.C., 1978.

Federal Power Commission. "Notice of Application for Preliminary Permit—Delaware River Development Company." 1956.

Feiveson, Harold A. "Conflict and Irresolution." In *Boundaries of Analysis,* edited by Harold A. Feiveson, Frank W. Sinden, and Robert H. Socolow. Cambridge, Mass.: Ballinger Publishing Co., 1976.

Francoline, Anthony J. "The Tocks Island National Recreation Area." Student report in the DRBC library, possibly Rutgers Dept. of City and Regional Planning. 5 April 1966.

Freeman, Smith, Edwin Mills, and David Kinsman. "Water Supply and Tocks Island Dam." Reprinted in *The Tocks Island Dam, A Preliminary Review: Papers in Support of a Free-Flowing Delaware River,* edited by George E. Schindler and Frank W. Sinden. Philadelphia: The Save the Delaware Coalition, 1973.

Gannett, Fleming, Cordry and Carpenter, Inc. *Report on Water Supply from the Upper Lehigh River for the City of Philadelphia.* Prepared for the Lehigh Coal and Navigation Co. Harrisburg, 1945.

————. *Report on Water Supply from the Upper Lehigh for the City of Philadelphia.* Prepared for the Lehigh Coal and Navigation Co. Harrisburg, 1946.

Gellhorn, Walter, and Frank P. Grad. "Opinion on the Constitutionality of the Proposed Delaware River Basin Compact." Prepared for the Delaware River Basin Advisory Committee. June 1960.

General Accounting Office. *Review of Tocks Island Reservoir Project: Comptroller General's Report to Chairman, Subcommittee on Public Works, Senate Committee on Appropriations.* Washington, D.C.: GAO, 1969.

————. *Problems in Land Acquisition for National Recreation Areas,* National Park Service. 1970.

Goodrich, Carter. *Government Promotion of American Canals and Railroads.* New York: Columbia Univ. Press, 1960.

Grossman, John. "Death of a Valley." *New Jersey Monthly* (May 1980).

Haefele, Mina. Statement at DRBC Tocks Island Public Information Meeting of 22 May 1975.

Hamilton, John D., and George Wharton Pepper. "A Memorandum and Supplementary Statements of Engineers Concerning the Upper Lehigh Water Supply." Submitted to the Mayor's Commission on Water Supply by the Lehigh Coal and Navigation Co. ca. 1946.

Harris-Dechant Associates. *Report, Diversion of the Waters of the Delaware River by the City of New York and Northern New Jersey.* Prepared for the Pennsylvania Dept. of Forests and Waters. Philadelphia, 1950.

Hazen, Whipple and Fuller. *Report on Water Resources of the State and Their Development.* Prepared for the New Jersey Dept. of Conservation and Development. New York, 1922.

Heacox, Cecil E. "Information on New York City Water Supply (Draft Report)." N.Y. Conservation Dept., 1949.

Heilich, Fredrick W. III. *The History of the Blairstown Railway.* Livingston, N.J.: Railroadians of America, 1981.

Hogarty, Richard A. *The Delaware River Drought Emergency.* Inter-University Case Program #107. Indianapolis: The Bobbs-Merrill Co., 1970.

Horton, Robert E. *A Program for Regulation and Conservation of the Delaware River.* Prepared for the Trenton Dept. of Public Works. Albany, 1929.

Hufschmidt, Maynard M. *The Interstate Commission on the Delaware River Basin—A Study of Its Role as a Planning Agency Engaged in the Political Process.* Cambridge, Mass., 1956. Mimeo.

————. *The Supreme Court and Interstate Water Problems—The Delaware Basin Example.* Cambridge, Mass., 1957. Mimeo.

"Hughes to Report to New York on Delaware Water Project." *Engineering News-Record* (26 January 1928): 170.

"Information Reports on Tocks Island Dam." *Congressional Record.* 20 May 1971: 4163–86.

Interstate Commission on the Delaware River Basin. *Annual Reports* or *Annual Reports of the Chairman.* 1937 to 1953.

————. *Articles of Organization.* 3 April 1936.

————. "Chronology of the Water Supply Problems of New York City, Philadelphia, and Northeastern New Jersey." ca. 1938.

————. "Chronological Summary of Abstracts of Acts of Assembly, Court Decisions and Other Pertinent Matters Regarding the Use of the Delaware River Watershed for Canals and Navigation." ca. 1938.

————. "Existing and Potential Water Project Dam Sites—Delaware River Above Trenton." Tabulated summary of pre-1946 dam sites with bibliography. ca. 1946.

————. Minutes of the Committee on Quantity. Various dates.

————. "Proceedings of Annual Meeting and Conference." Various dates.

Interstate Compact Between New Jersey and Pennsylvania Relating to Concurrent Jurisdic-

tion of the Delaware River. New Jersey Act of May 27, 1783, and Pennslyvania Act of September 20, 1783.

Interstate Water Management Recommendations of the Parties to the U.S. Supreme Court Decree of 1954 to the Delaware River Basin Commission Pursuant to Commission Resolution 78-20. Signed by Governors DuPont (Del.), Cuomo (N.Y.), Kean (N.J.), and Thornburgh (Pa.), and Mayor Koch (New York City). July 1984.

Jack McCormick and Associates. *An Appraisal of the Potential for Cultural Eutrophication of Tocks Island Lake.* Prepared for the U.S. Army Corps of Engineers. Devon, Pa., 1971.

Jersey Central Power and Light Co. *Yards Creek Pumped Storage Generating Station: Technical Description.* Morristown, ca. 1966.

Johnson, Lyndon B. "The Budget Message: The President's Message to the Congress Transmitting the Budget for Fiscal Year 1967." Dated 24 January 1966. In *Weekly Compilation of Presidential Documents.* 31 January 1966: 82.

―――. "The President's News Conference of January 17, 1967." In *Weekly Compilation of Presidential Documents.* 23 January 1967: 54.

―――. "The Budget Message: The President's Message to the Congress Transmitting the Budget for Fiscal Year 1969." Dated 19 January 1968. In *Weekly Compilation of Presidential Documents.* 5 February 1968.

Krass, Allan. "Floods and People." In *Boundaries of Analysis,* edited by Harold A. Feiveson, Frank W. Sinden, and Robert H. Socolow. Cambridge, Mass.: Ballinger Publishing Co., 1976.

Leahy, Gretchen. Statement at the 22 May 1975 DRBC Tocks Island Public Information Meeting.

Leahy, Gretchen, and Thomas Iezzi. Interview with author. Bristol, Pa., 1985.

Lehigh Coal and Navigation Company. "Testimony of the Lehigh Coal and Navigation Engineers as Presented to the Philadelphia Water Commission." 1946.

Lenni Lenape League. "The Case for Sunfish Pond." ca. 1968.

―――. "Fourth Annual Pilgrimage to Sunfish Pond." In newsletter. 10 April 1969.

―――. Letter to Jersey Central Power and Light Co., 9 August 1967.

"Major Rivers in North America." Pages 444–45 in *World Almanac and Book of Facts.* New York: Newspaper Enterprise Assoc., 1981.

Malcolm Pirnie Engineers, and Albright and Friel, Inc. *Report on the Utilization of the Waters of the Delaware River Basin.* Prepared for Incodel. New York, 1950.

Manning, Ralph W. *The Poultry Industry in the Upper Delaware River Basin.* West Trenton, N.J.: DRBC, 1971.

Matheson, Joan. Correspondence with author, 1982–85.

Menzies, Elizabeth G. C. *Before the Waters.* New Brunswick: Rutgers Univ. Press, 1966.

Moore, A. Harry. *Special Message of A. Harry Moore, Governor of New Jersey to the 163rd Legislature of New Jersey on the Delaware and Raritan Canal as a Source of Water Supply for New Jersey Together with Copy of Message to 162nd Legislature on the Same Subject.* Trenton, 1939.

Morrissey, Phillip. Testimony at House hearing concerning Tocks Island appropriations, 5 June 1969. In *Congressional Hearings, Public Works for Water and Power Development.* Washington, D.C., 1969.

Murdoch, Paxson, Kalish and Dilworth. "Legal Opinion—The Rights of the Lehigh Coal and Navigation Company in the Lehigh River and Its Tributaries." Prepared for the Philadelphia Water Commission. 1946.

Murray, W. S., and others. *A Superpower System for the Region Between Boston and Washington.* U.S. Geological Survey Professional Paper 123. Washington, D.C.: GPO, 1921.

National Park Service. *Master Plan, Delaware Water Gap National Recreation Area.* 1966.

Nealon, William J. "Opinion in Delaware Valley Conservation Association vs. Resor, Udall and Cassiday." U.S. District Court for the Middle District of Pennsylvania. 5 June 1967.

New Jersey Department of Conservation and Development. *Annual Report 1922–23.* ca. 1924.

New Jersey Department of Conservation and Development and the Delaware and Raritan Canal Commission. *Report on the Final Disposition of the Delaware and Raritan Canal in Accordance with Chapter 203, P.L. 1941.* Trenton, 1942.

New Jersey State Water Policy Commission. *Report of the State Water Commission for the Year 1908.* Patterson, 1909.

———. *Report of the State Water Supply Commission for the Year 1912.* Union Hill, 1912.

New Jersey Water Policy Commission. *Report of the Water Policy Commission, Part One: Dealing with the Proposed Compact as to the Waters of the Delaware River.* Trenton, 1926.

———. *Report of the Water Policy Commission, Part Two.* Trenton, 1926.

———. *Water Supply Problems of the Northern Metropolitan District, Special Report 1.* Newark, 1929.

———. *The South Branch Project: A High Level Water Supply for the Northern Metropolitan District, Special Report 3.* Newark, 1931.

New York City, Board of Water Supply. *Report of the Board of Water Supply to the Board of Estimate and Apportionment.* 1927. Reprinted 1937.

New York State Water Supply Commission. *Third Annual Report: Progress Report Under Chapter 569 of the Laws of 1907.* Albany, 1908.

———. *Fourth Annual Report.* 1909.

Northrop, Vernon D., and Paul N. Ylvisaker. "Staff paper on Delaware River Basin Development." Philadelphia Mayor's Office, 1955.

Pantzer, Kurt F. *Report of the Special Master Recommending Amended Decree.* Prepared for the U.S. Supreme Court. Indianapolis, 1954.

Pennsylvania Department of Highways. *Highway Impact Study, Delaware Water Gap National Recreation Area.* Harrisburg, 1966.

Pennsylvania Water Resources Committee. *Report of the Pennsylvania Water Resources Committee to Governor John S. Fine.* 1953.

Pitkin, Francis A. "Tocks Island in 1958." Speech given at Pocono Manor, Pa. In *Annual Report of the Chairman, 1957.* Philadelphia: Incodel, 1958.

Raymond and May Associates. *Preface to Planning: A Sketch Plan for the Tocks Region.* White Plains, N.Y., 1966.

Read, William. Interview with author. East Stroudsburg, Pa., 1984.

Reich, Michael. "Historical Currents." In *Boundaries of Analysis*, edited by Harold A. Feiveson, Frank W. Sinden, and Robert H. Socolow. Cambridge, Mass.: Ballinger Publishing Co., 1976.

Robert R. Nathan Associates. *Central Park in Megalopolis: The Potential Impact of the Delaware Water Gap National Recreation Area on Its Surrounding Communities.* Washington, D.C.: Communication Service Corp., 1966.

Roy F. Weston, Environmental Scientists and Engineers. *Tocks Island Region Environmental Study.* 5 vols. Prepared for DRBC. West Chester, Pa., 1970.

Schindler, George E., and Frank W. Sinden, eds. *The Tocks Island Dam, A Preliminary Review: Papers in Support of a Free-Flowing Delaware River.* Philadelphia: The Save the Delaware Coalition, 1973.

Schrader, Thomas F., and Robert H. Socolow. "Electric Power on the Delaware." In *Boundaries of Analysis,* edited by Harold A. Feiveson, Frank W. Sinden, and Robert H. Socolow. Cambridge, Mass.: Ballinger Publishing Co., 1976.

Shapp, Milton J. Letter from the Governor of Pennsylvania to Governor Brendan T. Byrne of New Jersey, 22 June 1978.

———. Letter from the Governor of Pennsylvania to Governor Hugh L. Carey of New York, 22 June 1978.

Sheldon, Walter J. *Tigers in the Rice: The Story of Vietnam from Ancient Past to Uncertain Future.* New York: The Macmillan Co., 1969.

Shukaitis, Nancy. Interview with author. Stroudsburg, Pa., 1982.

Smith, Douglas W. "A Walk with Casey Kays." *New Jersey Outdoors* (January 1985).

Snyder, Frank E., and Brian H. Guss. *The District: A History of the Philadelphia District, U.S. Army Corps of Engineers.* Philadelphia: U.S. Army, Corps of Engineers, Philadelphia District, 1974.

"Status of Delaware River Compact." *Engineering News-Record* (7 May 1925): 784.

Strandburg, William B. *Control of Salinity Intrusion in the Delaware River Estuary.* Prepared for DRBC. West Trenton, N.J.: DRBC, 1975.

"Summary of Minutes of Meeting of Representatives of the States of Delaware, New Jersey, New York and Pennsylvania and the Cities of New York and Philadelphia." Held in Princeton, 17 May 1955.

Thompson, Frank, Jr. Letter to Joe Evins, 1 March 1974.

———. Press release. 1 June 1967.

"Time Essay: How to Cut the U.S. Budget." *Time* (8 December 1967): 38.

Tippetts-Abbett-McCarthy-Stratton. *Survey of New Jersey Water Resources Development.* Prepared for the Legislative Commission on Water Supply. New York, 1955.

Tocks Island Regional Advisory Council. "Memo from TIRAC No. 1." 22 April 1966.

URS/Madigan-Praeger Inc. and Conklin and Rossant. *A Comprehensive Study of the Tocks Island Lake Project and Alternatives.* 6 vols. Prepared for the U.S. Army, Corps of Engineers. New York, 1975.

U.S. Army. Corps of Engineers. Philadelphia District. *Delaware Estuary Salinity Intrusion Study.* 1982.

———. Philadelphia District. *Delaware River Basin Report.* 11 vols. 1960. Reprinted as House Document 522, 1962.

———. Philadelphia District. *Design Memorandum No. 1: Site Selection.* 1965.

———. Philadelphia District. *Design Memorandum No. 1: Site Selection, Supplement No. 1.* 1967.

———. Philadelphia District. *Design Memorandum No. 3: General Design Memorandum.* Vol. 1. 1969.

———. Philadelphia District. *General Design Memorandum, Condensed Version.* 1971.

———. Philadelphia District. *Design Memorandum No. 6: Site Geology.* 1967.

———. Philadelphia District. *Design Memorandum: Spillway and Outlet Design.* 1972.

————. Philadelphia District. *Draft Environmental Impact Statement, Tocks Island Lake Project*. 1974.

————. Philadelphia District. *Environmental Impact Statement—Tocks Island Lake*. 1971.

————. *Information Bulletin—Delaware River Basin Study, Brief Summary of Water Needs and Tentative Plan of Development*. 1960.

————. Philadelphia District. *Preliminary Examination Report: Review of 308 Report on the Delaware River*. 1946.

————. Pittsburgh District. "Chronology of a Hurricane." Visitor brochure, n.d.

————. Pittsburgh District. "Fish of the Kinzua Dam and Allegheny Reservoir." Visitor brochure, n.d.

U.S. Congress. *A Legislative History of the Water Pollution Control Act Amendments of 1972*. Washington, D.C.: Library of Congress, 1973.

————. House. Committee on Government Operations, Commission on Organization of the Executive Branch of the Government. *Water Resources and Power Report, Part I*. (Hearings held in Mt. Pocono, Pa., on 21 and 22 September 1955.) 84th Cong., 1st sess., 1955.

————. House. Committee on Public Works. *Tocks Island Deauthorization: Hearings Before the Subcommittee on Water Resources, July 23 and 26, 1976*. 95th Cong., 2d sess., 1976.

————. House. *Delaware River 308 Report*. 73d Cong., 2d sess. H. Doc. 179. Washington, D.C.: GPO, 1934.

————. House. Resolution of the Committee on Rivers and Harbors. 24 January 1939.

————. House. Subcommittee on National Parks and Recreation. Committee on Interior and Insular Affairs. *Hearings on H.R. 89 and Related Bills*. 89th Cong., 1st sess., 1 March 1965 (Washington) and 22 April 1965 (East Stroudsburg).

————. Public Law 87-874, River and Harbor Act of 1962. 87th Cong., 2d sess., 23 October 1962.

————. Public Law 89-158, An Act to Authorize Establishment of the Delaware Water Gap National Recreation Area and for Other Purposes. 1 September 1965.

————. Public Law 90-542, Wild and Scenic Rivers Act. 2 October 1968.

————. The National Environmental Policy Act of 1969, Public Law 91-190. 1 January 1970.

————. Public Law 91-282, River Basin Monetary Authorization and Miscellaneous Civil Works Amendments Act of 1970. 19 June 1970.

————. Public Law 95 625, The National Parks and Recreation Act of 1978. 10 November 1978.

————. Rivers and Harbors Act of March 3, 1925.

————. Rivers and Harbors Act of January 21, 1927.

————. Senate. Committee on Energy and Natural Resources. *Hearings Before the Subcommittee on Parks and Recreation, May 5, 1978*. 95th Cong., 2d sess., 1978.

————. Senate Resolution. 13 April 1950. Directed review of 308 plan relationship to the Incodel plan.

————. Senate Resolution. 14 September 1955. Directed review of 308 plan relationship to 1955 flood.

————. Senate Resolution. 20 February 1956. Directed study of Tocks and Wallpack Bend dam sites.

————. Senate Resolution. 28 April 1958. Directed study of salt barrier dam.

U.S. Forest Service. Allegheny National Forest. *Update 1981: A Report on Your Forest and Its Resources.* Warren, Pa., ca. 1981.

U.S. Geological Survey. *Floods of August 1955 in Northeastern States.* Geological Survey Circular 377. Washington, D.C.: GPO, 1956.

————. *Report of the River Master of the Delaware River for the Period December 1, 1966–November 30, 1967.* Washington, D.C.: U.S. Geological Survey, 1968.

U.S. Supreme Court. The Delaware River Case: New Jersey versus New York. 283 U.S. 336 (Opinion) and 283 U.S. 805 (Decree). Includes 60 vols. of testimony. 25 May 1931.

————. The Delaware River Case: New Jersey versus New York. 347 U.S. 995 (Decree). 7 June 1954.

Vermeule, Cornelius C. *Report on Water Supply, Water Power, the Flow of Streams and Attendant Phenomena.* Prepared for the Geological Survey of New Jersey as Vol. 3 of the Final Report of the State Geologist. Trenton: John L. Murphy Publishing Co., 1894.

Voight, William, Jr. *The Susquehanna Compact, Guardian of the River's Future.* New Brunswick: Rutgers Univ. Press, 1972.

Wagner, Karl B. "Summary of Data Regarding Water Supply for Philadelphia." Prepared in connection with "General Studies of the Delaware River Drainage Area." Unpublished. Pa. Water and Power Resources Board, 1938.

"Water Crisis Coming in the East." *Engineering News-Record* (7 July 1983): 32.

Water Resources Association of the Delaware River Basin. "Appropriations Committee Testimony on Tocks Island Project. Memo from WRA/DRB." 7 July 1971.

————. "A Tocks Island Regional Council?" 1963.

————. "Congress Endorses Tocks Island Conservation Project. Memo from WRA/DRB." 1 September 1973.

————. "Statement before the Public Lands Subcommittee of the U.S. Senate Committee on Interior and Insular Affairs." 6 July 1964.

————. *The Facts about Tocks Island.* Parts 4 and 5 prepared by the U.S. Army, Corps of Engineers. Philadelphia: WRA/DRB, 1962.

————. *The Keystone Project: Tocks Island Revisited.* Philadelphia, 1971.

————. *Tocks Island and Outdoor Recreation for the Crowded East.* Philadelphia, 1964.

Water Resources Foundation for the Delaware River Basin. *Water for Recreation, Today and Tomorrow.* Philadelphia, 1959.

Weidner, Charles H. *Water for a City: A History of New York City's Problem from Beginning to the Delaware River System.* New Brunswick: Rutgers Univ. Press, 1974.

Whitall, W. Brinton. Interview with author. Princeton, N.J., 1985.

————. "The Delaware River Basin: Drought Management." *Environmental Forum* 3, no. 3 (July 1984): 34–39.

Whitman, Ezra B. *Report on the Effect of the Compact as to the Waters of the Delaware River upon the City of Trenton.* Baltimore, 1927.

Wilson, Malcolm (Governor of New York). Letter to Joe Evins, 2 May 1974. In Minutes of 22 May 1974 Meeting of DRBC.

Index